CMOS Projects and Experiments:
Fun with the 4093 Integrated Circuit

Other Books of Interest by Newnes

CMOS Projects and Experiments: Fun with the 4093 Integrated Circuit

Newton C. Braga

Newnes

Boston Oxford Auckland Johannesburg Melbourne New Delhi

Newnes is an imprint of Butterworth–Heinemann.

Copyright © 1999 by Butterworth–Heinemann

 A member of the Reed Elsevier group

 Recognizing the importance of preserving what has been written, Butterworth–Heinemann prints its books on acid-free paper whenever possible.

 Butterworth–Heinemann supports the efforts of American Forests and the Global ReLeaf program in its campaign for the betterment of trees, forests, and our environment.

ISBN 0-7506-7170-X

British Library Cataloguing-in-Publication Data
A catalogue record for this book is available from the British Library.

The publisher offers special discounts on bulk orders of this book.
For information, please contact:

Manager of Special Sales
Butterworth-Heinemann
225 Wildwood Avenue
Woburn, MA 01801-2041
Tel: 781-904-2500
Fax: 781-904-2620

For information on all Newnes publications available, contact our World Wide Web home page at:
http://www.newnespress.com

10 9 8 7 6 5 4 3 2 1

Printed in the United States of America

Contents

Preface

Some special components, such as the 555, 741, 567, 4017, and other popular ICs, can be used as the basis of a large assortment of electronics projects. The 4093 CMOS is one of these components. Using this versatile IC, the author compiled this special "one-component cookbook" that furnishes the experimenter, student, and technician with a large selection of practical circuits.

During the past several years, the author, as a contributor to U.S., European, and Latin American electronics magazines, has collected many electronic projects using the 4093 CMOS IC. Practical circuits, as well as ones that can teach you a great deal about electronics, have been selected and included in this book.

This book has a dual aim, as we have two kinds of electronics experimenters: (1) those who want to improve or expand their understanding of some other areas of interest, such as audio, radio, computers, instrumentation, security, or even games, and (2) those who want to gain an understanding of basic electronic circuits and the 4093 CMOS IC, which is the basis of this work.

We should also mention another kind of reader: the high school student who wants to use electronics in scientific explorations. Many projects described in this book can be used in scientific experiments or middle school science projects.

Most of the projects described here can stand alone as individual devices. However, wherever possible, the circuits have been designed so they can be ganged with one or more other projects. For example, many projects of audio effects and sound generators can be joined with the audio output stages outlined herein to drive several types of loads ranging from low-power piezoelectric transducers to high-power loudspeakers.

We also have simple projects that use only a few low-cost components. These can be assembled in a single evening, as opposed to the more complicated projects that employ several ICs, transistors, and other parts.

To make it easier for the reader to choose appropriate projects, each project title is labeled with a "P" to indicate that it has practical uses or an "E" to indicate that it is designed for the experimenter to teach him something about circuits or devices. You will find also projects with both marks (E and P), and these can be used for either purpose.

Chapter 1 describes, in simple terms, the 4093 IC itself. After this brief introduction, the remaining chapter provide 135 projects in a straightforward manner, grouped by general types. It begins with simpler projects and progresses to the more complex. This format should enable the reader to learn the theory of the device quickly and prepare him to use the circuits most advantageously.

The required electronic components are listed with each circuit diagram. Secondary parts as sockets, chassis, enclosures, miscellaneous hardware, and so on are not specified, since the reader is free to choose these non-critical items according to his preference and demands.

The manner in which the circuits work, and acceptable modifications, are explained in practical terms so the reader can acquire additional knowledge of practical electronics as he progresses through the book.

Although many of the projects (the practical ones) are fun to build exactly as they are discussed here, you may think of possible modifications. I recommend that you go ahead and modify the circuits to suit your personal ends. There is wide latitude for circuit modifications, and most of them will be of value to experimenters who want to see how things work, even though each project's primary value is for the builder who wants to produce a practical, functional item.

Because our projects utilize a wide range of power supply voltages, we have included several different regulated and unregulated power supply projects so that the reader doesn't run up an expensive bill for batteries.

The power supply requirements of each project have been tailored to one or more standard voltages that cam be easily attained using commonly available batteries or power supply transformers. The voltage, in most instances, will be similar to those for projects described in electronics magazines, so a power supply from this book can also be used for many projects from other sources.

As a whole, I believe this book will serve a definite purpose and fill a possible void in your repertoire of exciting electronic items.

Newton C. Braga

Sources of the 4093 IC

The 4093 integrated circuit can be purchased from many dealers. The reader can visit the site of Partminer (http://www.partminer.com) and download a program to find electronic parts in the Internet.

As this book went to print, typing "4093" in the search program turned up the following distributors.

- Allied Electronics
- CalSwitch
- Digi-Key
- Electro Sonic
- Gerber
- Avnet Electronics
- JACO
- Marshall
- Newark
- Nu Horizons
- Pioneer
- Reptron
- NetBuy
- Partminer Direct
- Powell Electronics

In addition, you can find sources for most components on the World Wide Web at http://www.chipcenter.com, which is a cooperative enterprise from Arrow Electronics, Inc.; Aspect Development, Inc.; Avnet, Inc.; and CMP Media, Inc.

Acknowledgments

This book started as a casual conversation I had with Edison de Santi, my great friend, who was the first to suggest that I collect all the projects I had devised for the 4093 integrated circuit and assemble them in a single book. A special note of thanks should also go to my wife, Neuza, who helped me by typing many of the originals, and to my son, Marcelo, who sacrificed some walks and fun while I was working on the book.

1

The 4093 CMOS IC

1.1 The CMOS Family

The 4093 CMOS integrated circuit is a member of a large group of compatible building blocks. With these blocks, we can build circuits that make simple decisions, so they are also classified as *digital logic circuits*. The CMOS group of integrated circuits is composed of several types of devices that, because they have the same input/output electrical characteristics, can be interconnected without the need of intermediate circuitry.

There are many digital logic families available as TTL (transistor–transistor logic), RTL (resistor–transistor logic), and others, but the 4093 IC is a member of the CMOS (complementary metal-oxide-silicon) family, which has some important advantages over other families. These benefits include very low cost; ultra-low, noncritical power requirements; and an ultra-high input impedance.

These features, plus the inherent advantages of CMOS ICs over other logic devices, permit the logic system designer/experimenter to achieve outstanding electrical performance, high reliability, and simplified circuitry in a wide variety of equipment designs.

Some CMOS features are as follows:

- All devices inputs are open circuits and thus are easy to drive.
- They will run on ultra-low operating currents, particularly at low frequencies.
- They have high noise immunity—typically, 45 percent of supply voltage.
- They operate in a wide voltage range: 3 to 15 V (A-series devices) and 3 to 18 V (B-series devices).
- The inputs are fully protected.
- They feature large unloaded output swing: the output goes from ground potential (0 V) to the positive supply.
- They generate very little noise in power supply lines.

The manner in which CMOS integrated circuits are fabricated and the characteristics of all families can be found in several specialized books, normally classified as "digital electronics books," and in articles published in electronics magazines. If the reader wants more information about CMOS family, and digital electronics in general before to starting with our projects, a brief look in such sources might be valuable.

1.2 The 4093 IC

The 4093 CMOS integrated circuit is formed by four two-input NAND Schmitt triggers in a 14-pin package. All four positive-logic NAND gates may be used independently (Fig. 1.1).

In Fig. 1.1, we show the package used for this IC and also the NAND Schmitt trigger symbol. In terms of logic, the action of the circuit is the same as the action of the common NAND gate shown in Fig. 1.2.

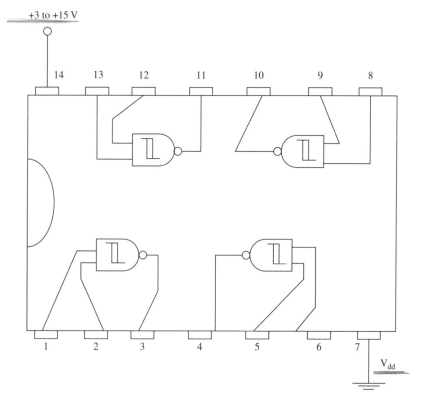

Figure 1.1 4093 CMOS IC, functional diagram. It is available in 14-pin dual in-line (DIP) packages.

Figure 1.2 NAND gate symbol and its truth table.

The output logic level depends on the input logic level. A zero appears at its outputs if both inputs are ones, and a one appears at its output if either or both inputs are zeros.

We should remind the reader that a zero (or a *low* logic level) is given by a 0-volt voltage, and a one (or *high* logic level) is given by the positive supply voltage—also called $+V_{cc}$ or V_{dd}.

For the 4093 IC and all devices of CMOS logic family, $+V_{cc}$, or the positive supply, can range from +3 to +15 V (A-series devices) or +3 to +18 V (B-series devices). For circuits that operate in a linear mode over a portion of the voltage range, such as RC or crystal oscillators, a minimum supply voltage of 4 V is recommended.

The difference between a common NAND gate and a Schmitt NAND gate is the *snap* action with hysteresis that it provides, also called *dead band*. Let's explain:

In Figure 1.3, we show the transfer characteristic of a 4093 ICs Schmitt trigger. The general shape of this characteristic is the same for all values of V_{dd} (positive supply) until V_p is reached.

At this point, the output goes low (0 V) and remains low as the input voltage is raised to V_{dd}. If the input voltage now is reduced, the output goes high and remains high as the input voltage is reduced to zero. The hysteresis is the difference between V_p and V_n, which is typically 0.6 V for a 5.0 Vs power supply.

Figure 1.4 shows the input/output characteristic of a 4093 device. Note that Schmitt triggers are bistable circuits that are driven by their inputs. They are useful for squaring up slowly rising or noisy inputs, contact debouncing, and oscillators. We can conclude that the 4093 offers unlimited possibilities for the experimenter, and in the following pages we show many of them. The reader, having a large imagination, can explore the basic circuits and get much more from them.

1.3 Basic Configurations

1.3.1 NAND Schmitt Trigger

Of course, the basic internal circuit of the 4093 is designed to be a NAND gate and therefore to operate as a logic unit that makes some simple decisions given by the table in Fig. 1.5. Accordingly, the basic application of this device is as a NAND gate.

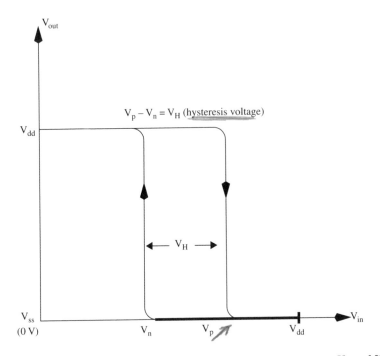

Figure 1.3   Transfer characteristic of the 4093. The difference between V_p and V_n is called *hysteresis voltage*.

$10^{12}\,\Omega$

The ultra-high input impedance of the device, typically $10^{12}\,\Omega$, allows the utilization of large input resistors in a circuit such as the one given in Fig. 1.5.

As we also see in that figure, the input voltage determines the output voltage in four possible conditions. We will use this basic configuration in several projects.

The maximum output currents of each gate depends on the power-supply voltage, and at $25°\,C$ they are as follows:

V_{dd} (V)	Max. current (sink and source) (mA)
5 V	1.0
10 V	2.6
15 V	6.8

1 → 7 mA

1.3.2   Inverter

The simplest possible logic block has one input and one output. We can make this simple block with the 4093 Schmitt triggers by wiring one

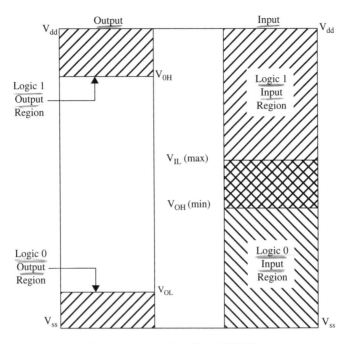

Figure 1.4   Input and output characteristics of the 4093 IC.

of the two inputs to the positive supply (V_{dd}) or wiring the two inputs together as shown in Fig. 1.6.

The truth table in Fig. 1.6 shows that, when the input is a one, the output is a zero, and vice versa. The block *inverts* the logic level applied to its input and so is called an *inverter*.

Inverters are used to generate the complement of a logic signal or to change the definition of a logic signal from positive to negative and back again (remember that in digital logic, the complement of 0 is 1, and the complement of 1 is 0).

To indicate that the block changes the signal, we use a small circle in its symbol output. In Fig. 1.7, we show how to use this circuit as a rise-time enhancer. Extremely slow signals or ultra-low frequency sinewaves can be converted to fast-rise outputs in this way.

The same configuration can be used to eliminate noise on a system input, as shown in Fig. 1.8. Note that, for complete rejection, the noise amplitude can be less than the dead band.

Contact debouncers or *conditioners* are shown in Fig. 1.9, based on a 4093 Schmitt trigger. The contacts debouncers are conceived for normally open and normally closed single switches, and these circuits operate by discharging or recharging a RC network. For best performance, the RC network time constant must exceeds three times the

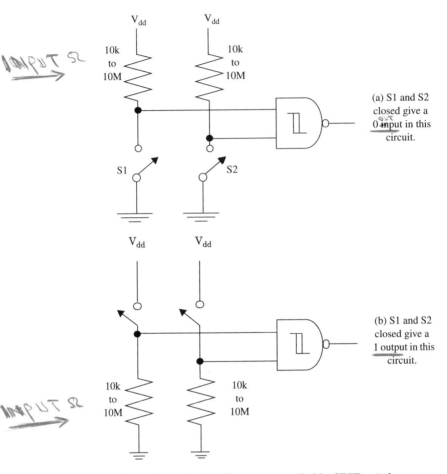

INPUT Ω

INPUT Ω

(a) S1 and S2 closed give a 0 input in this circuit.

(b) S1 and S2 closed give a 1 output in this circuit.

Figure 1.5   The 4093 used as NAND gates controlled by SPST switches.

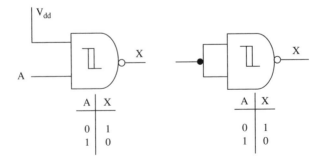

A	X
0	1
1	0

A	X
0	1
1	0

Figure 1.6   The 4093 wired as an inverter.

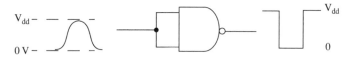

Figure 1.7   The 4093 wired as an inverter.

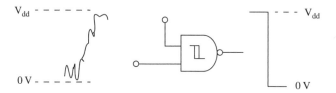

Figure 1.8   The 4093 as a noise eliminator.

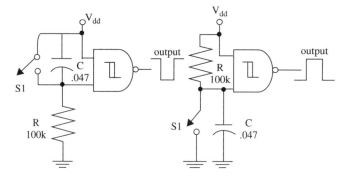

Figure 1.9   Normally open and normally closed contact debouncers.

worst-case bounce time. Depending on the power-supply voltage, the 4093 will operate from dc to 1 MHz in this application.

1.3.3   Oscillator

Stable oscillators may be constructed easily using the 4093 NAND gates. Figure 1.10 shows the basic circuit and its waveforms. It works as follows:

Before power is applied, input and outputs are at the ground potential (0 V, or logic level 0) and capacitor C is discharged. At power on, the output goes to high (V_{dd}, or logic level 1), and C charges through R until V_p is reached. The output then goes low. C is now discharged through R until V_n is reached. The output then goes high and charges C toward V_p through R. Therefore, we conclude that input alternatively swings between V_p and V_n as the output goes high and low.

OSCILLATOR *(handwritten)*

OUTPUT → *(handwritten)*

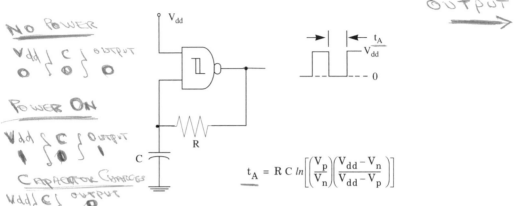

Handwritten notes (left margin):

NO POWER

Vdd	C	OUTPUT
0	0	0

POWER ON

Vdd	C	OUTPUT
1	0	1

CAPACITOR CHARGES

Vdd	C	OUTPUT
1	1	0

$$t_A = R\,C\,ln\left[\left(\frac{V_p}{V_n}\right)\left(\frac{V_{dd} - V_n}{V_{dd} - V_p}\right)\right]$$

Figure 1.10 Basic NAND-gate oscillator with the output waveform. A period of one oscillation is given by the formula.

An important advantage of this circuit that the oscillator is self-
starting at power on. We also see that the basic oscillator consists of an
amplifier (4093 gate) and a feedback network. For oscillation to occur,
the gain of the amplifier × attenuation of the feedback network must
be greater than 1. The frequency stability of an oscillator such as this
is primarily dependent on the phase-changing properties of the feedback network. For the 4093, we have good performance in frequencies
as high as 1 MHz, and the RC network values are given in the figure.

1.3.4 Monostable Multivibrator

The monostable multivibrator has one stable state and one unstable
state. It remains in the stable state until it is triggered. When triggered, it will be placed temporary in the other state. The multivibrator
will remain in that state during a time delay, after which it snaps back
to its original condition.

We can use monostable multivibrators to detect leading and trailing
edges of waveforms, in medium-accuracy generators, and also in time-
delay generators.

Each gate of a 4093 IC can be used as monostable multivibrator.
There are several ways to create a monostable multivibrator with the
gates as shown in Fig. 1.11.

In Fig. 1.11a, we show a positive-edge responding monostable multivibrator. In this configuration, the output goes to 0 during a definite
time delay given by the RC network, and it swings back to 1 when a
positive transition occurs in its input.

In Fig. 1.11b, a negative-edge responding configuration is shown.
The output swings from zero to one during a definite time delay when
the input voltage falls from one (+V_{cc}) to zero (0 V).

MONOSTABLE MULTIVIBRATOR

INPUT

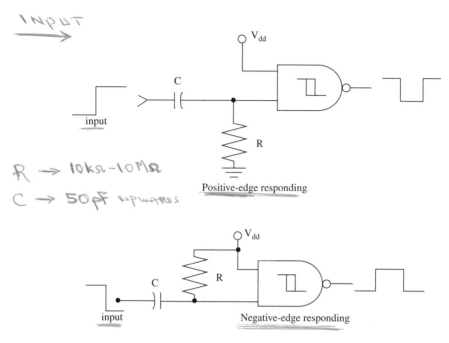

R → 10kΩ – 10MΩ

C → 50pF UPWARDS

Figure 1.11   Monostable circuits with the 4093.

Resistor values in the RC network can go from 10 kΩ up to 10 MΩ, and the capacitor can range from 50 pF on up.

Time delay as function of component values in the RC network is given in the chart shown in Fig. 1.12. Some usage limits should be observed for this kind of monostable multivibrator:

- The input voltage must stay in the post-trigger state for a period of time longer than the *on time*.
- The input transition must be noise free.
- The circuit may not be retriggered until after full recovery.
- The input must return into the pretrigger state long enough to let the RC network recover fully.

As, really, the circuits shown are edge detectors, they also can be called *half monostable*.

1.3.5 Set-Reset Flip-Flop

In Fig. 1.13, we show how two NAND gates of a 4093 can be wired to form a set-reset flip-flop. The circuit is a pulse-triggered flip-flop that needs a negative-moving pulse to be triggered. This circuit operates as follows.

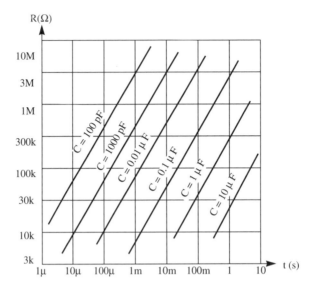

Figure 1.12 Component values for monostable operation.

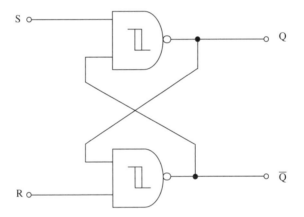

Figure 1.13 Set-reset flip-flop using two gates of a 4093 IC.

As we can see, the circuit has two outputs: a normal output, called Q, and a inverted output called $\overline{Q}$. When one output is 1, the other necessarily will be a 0, and vice-versa, because they are complementary.

The circuit has also two inputs named S (set) and R (reset), as shown in the figure, where the trigger signals are applied. R input is wired to Q output, and S input is wired to $\overline{Q}$ output, performing a closed-loop for the digital signals.

When a negative-moving trigger pulse is applied to S output, the output Q swings to the 1 state. As this output is wired to input R, the 1 state causes output $\overline{Q}$ to fall to a 0 level. But Q output is also wired to input S, causing a feedback that causes its output to remain at 1 even after the trigger pulse has disappeared. To trigger the multivibrator again, changing the output states, we should apply a negative-moving pulse to input R. This pulse causes the output to go 1 and, as this input is wired to input R, the trigger pulse also causes the output Q to go to a 0 level.

A zero in this output Q goes to input S and also, after the trigger pulse disappears, the outputs remain in their states. Note that the circuit has two stable states, and we only can change these states with set or reset (R or S) negative-moving pulses applied to its inputs. A manually triggered flip-flop can be constructed by wiring switches to the set-reset inputs as shown in Fig. 1.14.

The 4093 high input impedance allows the use of a large range of resistor values in this application. They typically can range from 1 kΩ to 10 MΩ. Resistors are used to apply a high logic level at the inputs when the switches are open. Without the resistors, we will have an indefinite state at the inputs when the switches are off, and this could cause erratic operation of the circuit.

With high-value resistors in this circuit we can get a touch-triggered bistable as shown in Fig. 1.15. Many projects in this book use that configuration as the starting point. We can obtain different performances for this basic circuit using other gates as inverters and obtain positive-moving bistables.

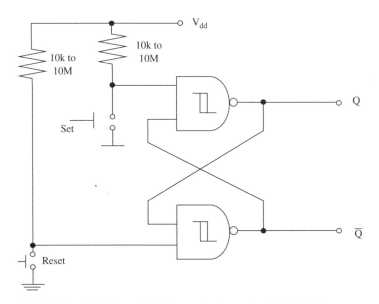

Figure 1.14 Manually triggered flip-flop using two gates of a 4093 IC.

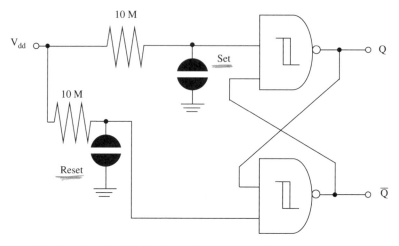

Figure 1.15   Touch-controlled flip-flop using two gates of a 4093.

1.3.6   Two-Gate Oscillator

Two NAND Schmitt triggers of a 4093 can create a two-gate oscillator, as shown in Fig. 1.16. This circuit generates a square wave with excellent thermal stability and operates in a large range of frequencies. With the values shown in Fig. 1.16, it operates at about 1 kHz.

In practice, C1 must be a nonpolarized capacitor and can have any value from 50 pF to some microfarads. R1 can have any value in the range between 4.7 kΩ and 20 MΩ. As in other applications, for a variable frequency operation, wire a fixed and a variable resistor in series in the R1 position.

The voltage in the junction of C1 and R1 is prevented from swing below zero or above the positive supply level by the built-in clamping diodes at the 4093 inputs, which causes the operating frequency of the circuit to be somewhat dependent on the power supply voltage. A sim-

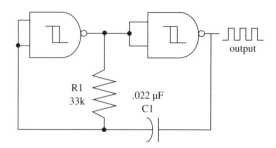

Figure 1.16   Two-gate oscillator running at 1 kHz.

ple rule shows that the frequency falls about 0.1 percent for each 1 percent rise in supply voltage.

The deficiencies of the basic circuit are minimized by using a *compensated* configuration as shown in Fig. 1.17. In this circuit, R2 is wired in series with the first gate input, and it allows the R1–C1 junction to swing freely below zero and above the positive supply. Values for this component typically range between 2 and 10 times R1. In this circuit, frequency varies only 0.5 percent when the supply voltage rises from 5 to 15 V. Other components can be added to the basic circuit for better performance, as shown in Fig. 1.18. Diodes can be

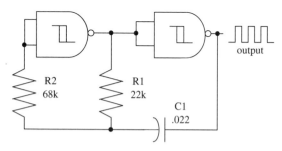

Figure 1.17   Improved two-gate oscillator with the 4093 IC.

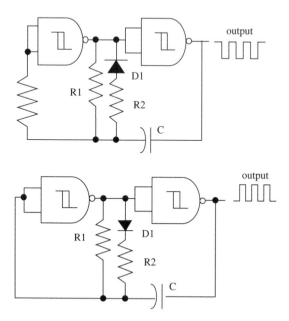

Figure 1.18   Nonsymmetrical waveforms can be generated with these circuits. R1 and R2 determine the duty cycle of the output signal.

used to give a nonsymmetrical output, as we will see in some practical applications.

As we have a signal traveling through two gates rather than just one, this circuit is slower than the one-gate oscillator. The maximum practical operation frequency for a 6 V power supply is about 500 kHz.

1.3.7 Three-Gate Oscillator

Three of the four gates of a 4093 can be used to make an oscillator with very clean output waveform, as shown in Fig. 1.19. Two gates as inverters wired in series give a high-gain noninverting stage and, due the electrical characteristic of this configuration, a fast rise time and fall time in the outputs are produced, which are directly suitable for use as clock generators. All the variations we have seen in other oscillators can be subjected to this basic design, as shown in Fig. 1.20.

1.3.8 Controlled Oscillators

We can use one input of each gate of a 4093 to control all oscillator operations. Therefore, all the oscillators can be modified easily for gated operation, so they can be turned on and off via an external signal.

In Fig. 1.21 we show a one-gate controlled oscillator. This circuit has a normally low output (0 V) and is gated by a high input signal.

Figure 1.22 shows a controlled two-gate oscillator. This oscillator is turned on by a high input signal.

Controlled three-gate oscillators are shown in Fig. 1.23. The circuit shown in Fig. 1.23a has a normally low output and is gated by a high logic level applied to the gate.

The circuit shown in Fig. 1.23b has a normally high output and also is gated by a high logic input.

In Fig. 1.23c, we have a controlled three-gate oscillator gated by a low logic level input.

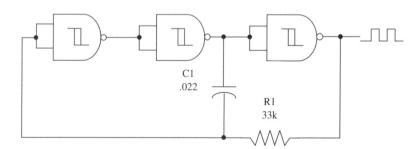

Figure 1.19   A three-gate oscillator using the 4093 IC.

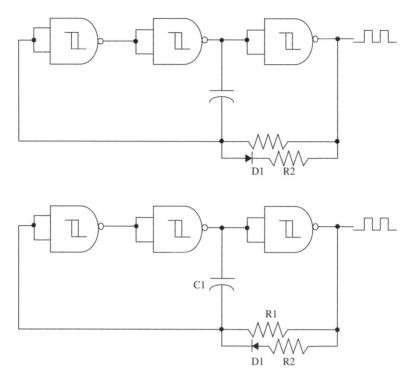

Figure 1.20 Asymmetric outputs produced by diodes.

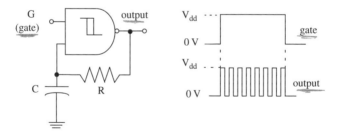

Figure 1.21 Gated one-gate oscillator using the 4093 IC.

1.3.9 Digital Amplifiers and Buffers

A high-impedance digital signal applied to the input of a 4093 gate, wired as an inverter results in a low-impedance digital signal. The output signal has the same waveform but this is valid only with digital signals as the square waveforms shown in Fig. 1.24.

We can consider this configuration to be a unity-gain voltage amplifier, but a high-current-gain amplifier. This circuit can be used in sev-

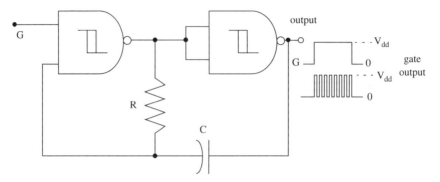

Figure 1.22 Gated two-gate oscillator using the 4093 IC.

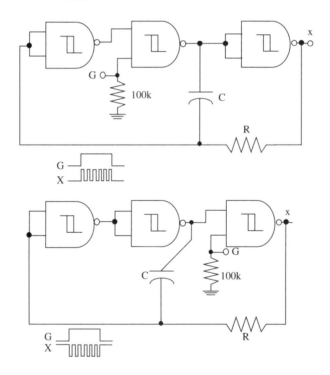

Figure 1.23 Three-gate oscillators using the 4093.

eral practical application, as will see in the following pages. For instance, a digital amplifier can be used to drive powerful loads with weak digital signals, as shown in Fig. 1.25.

Note that this an inverter configuration, so the output is a 1 when the input is a 0, and vice versa. A *phase-shift* is added to the amplified signal.

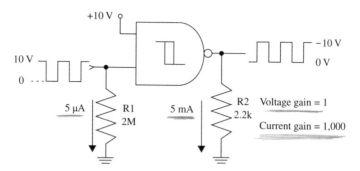

Figure 1.24   Digital amplifier. R1 and R2 determine the current gain.

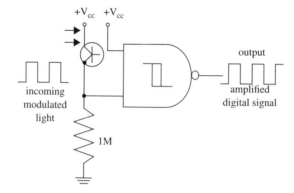

Figure 1.25   Digital amplifier using a photo-transistor wired to the input.

This circuit can also be used to isolate an oscillator or other input-circuit from an output, as shown in Fig. 1.26.

If we connect a heavy load to the output, as shown in Fig. 1.27a, the oscillator will show erratic operation. This problem can be solved with a buffer, as shown in Fig. 1.27b.

1.3.10 Outputs

What a 4093 IC output can drive is an important consideration to the experimenter who wants to make his own project. CMOS integrated circuits such as the 4093 have low output capability. Even with a 10 V power supply, each gate cannot source or sink more than a few milli-amperes, as suggested in Fig. 1.28.

Only light loads as LEDs, piezoelectric transducers, crystal ear-phones, can be driven directly by these outputs. To get more power from CMOS integrated circuits, we have to employ some special cir-cuits. For example, if your project does not use all the 4093s gates, you

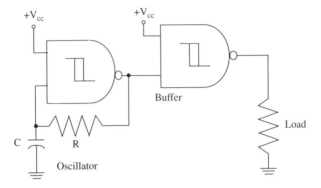

Figure 1.26 A 4093's gate used as a buffer.

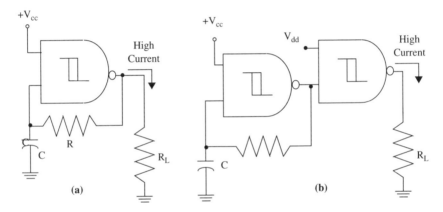

Figure 1.27 A gate used to isolate the load from the oscillator.

can connect the unoccupied gates to give extra power to the load. Connect the gates in parallel as shown in Fig. 1.29.

We have twice the maximum current using two gates, and with three gates we can get three times the maximum current. Using a 10 V power supply, you can source or sink about 10 mA with these circuits.

More power can be applied to the loads with a push-pull stage, as shown in Fig. 1.30. Two or four gates can be used to drive audio loads with currents up to some tens of milliamperes.

However, if you want to drive more powerful loads, such as loudspeakers, relays, lamps, and so on, you need much more power, and this implies that special output stages be used. Transistors are suitable components for this task.

Figure 1.31 shows two output stages that can be used to drive loads with current requirements up to 100 mA. The circuit shown in

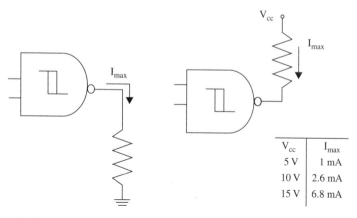

V_{cc}	I_{max}
5 V	1 mA
10 V	2.6 mA
15 V	6.8 mA

Figure 1.28   Current drain on the source depends on the power supply voltage.

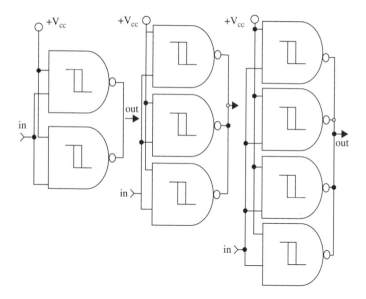

Figure 1.29   Gates can be wired in parallel to drive powerful loads.

Fig. 1.31a uses an NPN transistor. The transistor in this circuit is biased to conduct when the gate is at 1 logic level. The circuit shown in Fig. 1.31b uses a PNP transistor and is biased to conduct with a 0 in the gate output.

With small-signal transistors and audio transducers, we can get powers ranging from milliwatts to hundreds of milliwatts. The load impedance can range from 4 to 80 Ω, depending on the power supply voltage and the application.

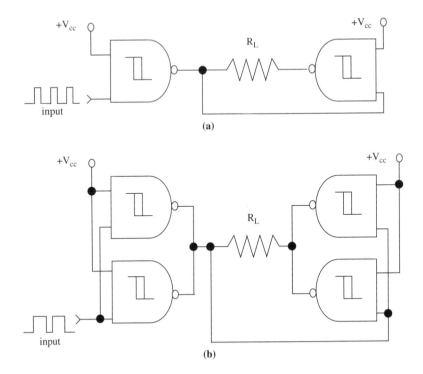

Figure 1.30   Push-pull or *bridge* stages to drive high-power loads.

With a 12 V power supply and low-impedance loads, medium power transistors should be used, such as the 2N2218, BD135, and TIP31 (all NPN), or the BD136 and TIP32 (all PNP). These transistors must be mounted on heatsinks. The output currents can be raised to 1 A in these circuits.

To get much more power without loading the IC's output, we can also use a Darlington transistor. Darlington transistors with current gains ranging from 800 to 10,000 are common. The circuits are shown in Fig. 1.32. TIP120, TIP121, and TIP122 (NPN), and TIP115, TIP116, and TIP117 (PNP) devices are suitable for this application. These transistors can supply loads with currents up to 3 A and must be mounted on heatsinks.

A Darlington pair can be improvised with two common transistors, as shown in Fig. 1.33. The total current gain for this configuration is given by the first transistor gain × the second transistor gain. Typical base-resistor values are in the range of 1 to 47kΩ.

Complementary pairs (an NPN and a PNP transistor with the same current-gain and other characteristics) can also be used to drive a low-frequency load with high power, as we show in Fig. 1.34. Figure 1.34a

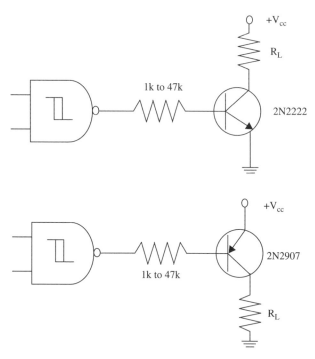

Figure 1.31   NPN and PNP transistors driving loads from logic signals of a 4093.

shows a *low-power* complementary output stage with small signal transistors supplying some hundreds of milliwatts to a loudspeaker. More power is given by the output stage shown in Fig. 1.34b. Medium-power transistors are used with a 12 V power-supply.

The 4093 IC's outputs can also drive SCRs and triacs. Sensitive SCRs, such as the TIC106, can be triggered directly with the output signals. In the circuit suggested in Fig. 1.35, the SCR goes on with a high level in the CMOS output. Triacs and less sensitive SCRs can be triggered with boosted signals, as shown in Fig. 1.36. Transistors are used to get more trigger current. The resistor value depends on the trigger requirements and the applications intended for the circuit.

Finally, you can also use power FETs to get high power outputs from the 4093 IC. A configuration used to drive dc loads with current drain up to 10 A is shown in Fig. 1.37.

The current drain of this configuration depends upon the power FET characteristics. Power FETs rated to drain currents up to 10 A are common and can be used in many applications described in this book. Types such as IRF640, IRF720, and others of the series IRF can replace the TIP Darlington and common transistors in almost all of the projects described in this book to get better performance.

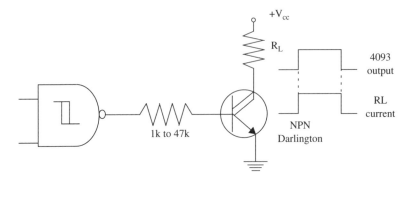

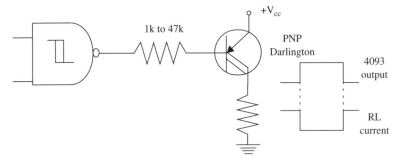

Figure 1.32   Darlington transistors can be used to drive heavy loads.

1.4 Power Supplies

Power requirements for the projects depend on their applications, typically ranging from a few milliamperes to one or two amperes. Low-power projects can be supplied with AA cells, 9 V batteries, or even 12 V wet cells. Higher-power projects should be supplied with appropriate devices using transformers, rectifiers, and filters.

Some simple power supplies are suggested in the this section for the experimenter who does not need to use expensive cells in the projects.

1.4.1   Simplest Power Supply

Currents up to 500 mA can be supplied by the circuit shown in Fig. 1.38. You can use this power supply to power the projects with current ratings up to 500 mA and voltages rates from 6 to 9 V.

Critical projects should not be supplied with this circuit, as it has a nonregulated output.

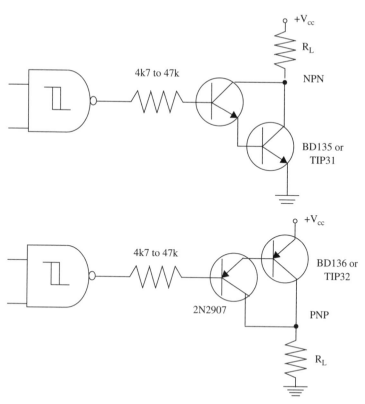

Figure 1.33 "Home-made" Darlington transistors used to increase power for driving heavy loads.

Parts List: Simplest Power Supply

Transformer	Primary, 117 Vac Secondary, 6 Vac, 500 mA
D1, D2, D3, D4	50 V × 1 A, 1N4002 or equivalent silicon rectifier diodes
C1	1,000 μF × 16 WVdc electrolytic capacitor
S1	SPST switch
F1	500 mA fuse

Figure 1.39 shows an exploded view of the power supply that can be housed in a small plastic box. Banana plugs and jacks are used to connect the power supply to the external circuit.

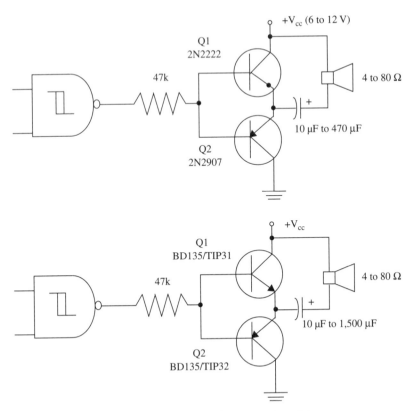

Figure 1.34 Power output stages with complementary transistors.

1.4.2 Regulated Power Supply

This circuit can be used to provide 6 and 12 V x 1A regulated outputs to experimental circuits. The output voltage depends on the IC used. The integrated circuit must be mounted on a heatsink.

Parts List: Regulated Power Supply

IC1	7806 (6 V) or 7812 (12 V) positive three-terminal regulator
T1	Primary: 117 Vac Secondary, 12 V or 15 V × 1 A
D1, D2, D3, D4	50 V × 1A, 1N4002 or equivalent silicon rectifier diodes
C1	1,000 µF × 25 WVdc electrolytic capacitor
C2	100 µF × 12 WVdc electrolytic capacitor
F1	500 mA fuse
S1	SPST switch

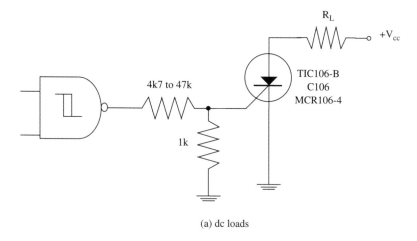

(a) dc loads

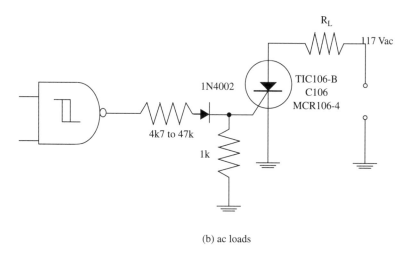

(b) ac loads

Figure 1.35   Triggering sensitive SCRs from a 4093 output.

Figure 40 shows the complete project assembled in a small plastic box.

1.4.3   Boosted Power Supply

High current, ranging from 0 to 5 A, can be supplied with the circuit shown in Fig. 1.41. Output voltage depends on the IC. For 6 V output, use a 7806, and for 12 V, use a 7812 IC.

Transistor Q1 must be mounted on a heatsink, and R1 in conjunction with the base-emitter voltage (V_{BE}) of the transistor determines when the pass transistor begins conducting. This circuit is not short-circuit proof.

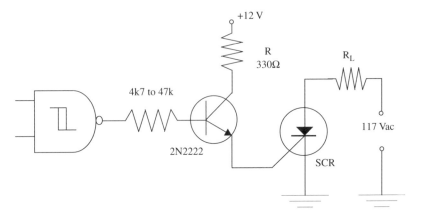

(a) More current to trigger SCRs or triacs can be obtained from this configuration.

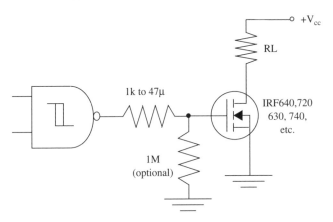

(b) A power FET can replace a Darlington transistor for driving high-power loads.

Figure 1.36   More configurations for driving loads with a 4093.

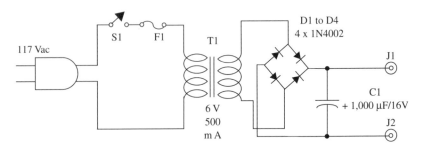

Figure 1.37   Simple unregulated power supply.

Parts List: Boosted Power Supply

IC1	7806 or 7812 IC, according to output voltage
Q1	MJ2955 power PNP transistor
D1–D4	50 V × 5 A silicon rectifier diodes
R1	56 Ω, 5 W wire wound resistor

T1	Transformer:
	  Primary, 117 Vac
	  Secondary, 12 or 15 Vac, 5 A

C1	2,200 µF, 25 WVdc electrolytic capacitor
C2	1.0 µF, 25 WVdc electrolytic capacitor
C3	0.1 µF ceramic or metal film capacitor

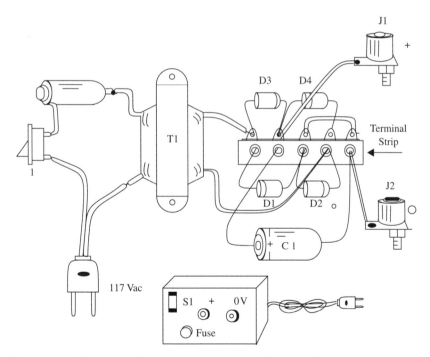

Figure 1.38 Component placement using a terminal strip as chassis.

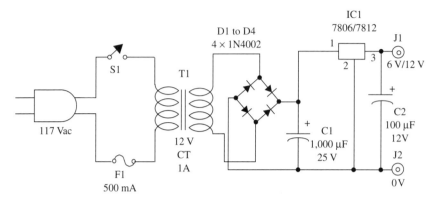

Figure 1.39 Regulated power supply with current output up to 1 A.

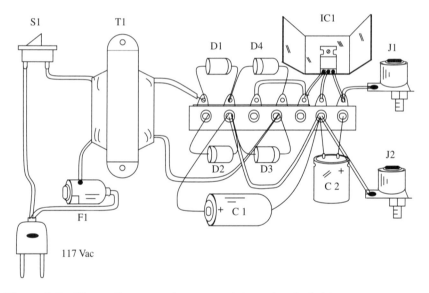

Figure 1.40 The small components are mounted on a terminal strip.

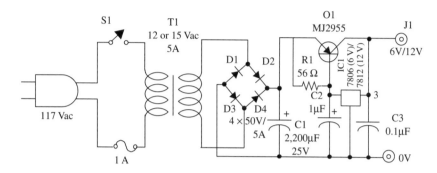

Figure 1.41   Boosted power supply. Q1 must be mounted on a heatsink, as is IC1.

2

Audio and RF Projects

The projects described in this chapter use 4093 ICs as audio frequency (AF) and high frequency (RF) oscillators. These circuits can drive loads as piezoelectric transducers, small loudspeakers, earphones, and so on. Based on the information given in the previous Chapter 1, the reader can easily alter the output stages to drive other loads requiring other power levels.

All of the projects can be mounted on solderless boards (experimental versions), printed circuit boards, or universal printed circuit boards (builder's versions).

Project 1: Simple Audio Oscillator I (E)

This low-power experimental oscillator can generate audible signals in the range of 100 Hz to 1.2 kHz, driving a small piezoelectric transducer or a crystal earphone. The project can be powered from four AA cells (6 V), a 9 V battery, or a 12 V power supply or battery. The circuit drains only a few milliamperes, which can extend the batteries' life to many days if they are used to power the circuit.

Parts List: Simple Audio Oscillator I

IC1	4093 CMOS integrated circuit
BZ	Piezoelectric transducer, Radio Shack 273-073 or equivalent (a crystal earphone also can be used)
R1	100,000 Ω potentiometer
R2	10,000 Ω, 1/4 W, 5% resistor
C1	0.022 or 0.033 µF ceramic or metal film capacitor
C2	100 µF, 16 WVDC (working voltage, dc) electrolytic capacitor

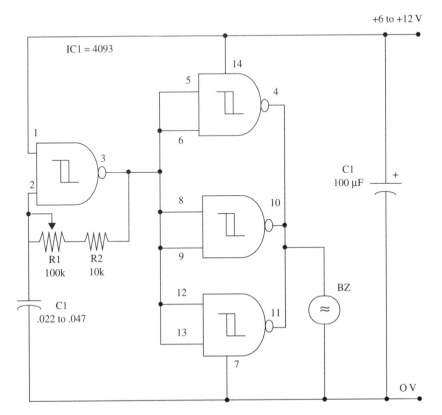

Figure 2.1 Simple Audio Oscillator. BZ can be any piezoelectric transducer or an earphone (high-impedance type).

Potentiometer R1 adjusts the frequency and can be altered in a large range of values. Potentiometers up to 1 MΩ can be used, changing the frequency range lower limit to about 10 Hz.

C1 can also be altered, and values between 0.01 µF and 0.1 µF are suitable. Large C1 values will give lower frequencies. This circuit can be used as a part for alarms, games, toys, and to learn a great deal about the 4093.

The output waveform is square, and frequencies between 1,500 and 3,000 Hz will provide more audio power with the recommended Radio Shack transducer.

Project 2: Simple Audio Oscillator II (E, P)

This is a "high-power" version of Simple Audio Oscillator I, with a transistorized power output stage driving a small low-impedance loudspeaker. With a 12 V power supply, you can get some hundreds of milliwatts of continuous sound from this circuit. Operational frequency is adjusted by potentiometer R1.

As in the previous project, you can alter the frequency range by changing R1 and/or C1. Values between 100 kΩ and 1 MΩ for R1, and 0.01 to 0.22 µF for C1, can be used in this project.

The schematic diagram of Audio Oscillator II is given in Fig. 2.2.

Parts List: Audio Oscillator II

IC1	4093 CMOS integrated circuit
Q1	2N2222 (6 V), 2N2218 (9 V), BD135 (9 V), or TIP31 (12 V) NPN transistor (see text)
SPKR	4 or 8 Ω small loudspeaker
R1	100,000 Ω potentiometer (linear or logarithmic)
R2	10,000 Ω, 1/4 W, 5% resistor
R3	4,700 Ω, 1/4 W, 5% resistor
C1	0.022 µF metal film or ceramic capacitor
C2	100 µF, 16 WVDC electrolytic capacitor

Transistor Q1 depends on the power supply voltage. In the 9 V and 12 V versions, the transistor must be mounted on a heatsink. You can also use a power FET—any *IRF* type with drain current rated to 2 A or more.

The circuit can also be used as part of alarms, toys, games, and to teach the reader a great deal about audio oscillators using the 4093.

Current requirements for this circuit depend on the power supply voltage and the loudspeaker's impedance. Typically, they range from 10 to about 500 mA. With power FET and supplies between 12 and 15 V, the circuit can drain up to 2 A.

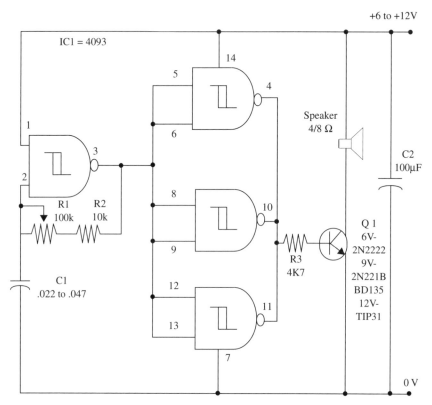

Figure 2.2 Audio Oscillator II. Output transistor depends on the power supply voltage.

Project 3: Signal Injector (E, P)

If you are interested in servicing on your audio equipment, you will undoubtedly find this square wave generator to be a handy tool. But you can also use this circuit in RF stages, as in AM/FM receivers, since the oscillator harmonics are strong in frequencies as high as 100 MHz.

In this arrangement, frequency is determined by capacitor C1 and resistor R1 values. In our circuit, the oscillator runs at 1 kHz.

The output signal waveform is square, and the output swings the full power supply voltage, which can be anything between 6 and 15 V. (A 9 V battery is recommended as the power supply for this project).

Supply current is typically 10 μA, and the other three nonoscillating gates are used as digital amplifiers and to buffer the oscillator. The frequency can be altered by selection of the resistor and/or capacitor value in the oscillator stage.

Figure 2.3 shows the complete schematic diagram of the signal injector.

Parts List: Signal Injector

IC1	4093 CMOS integrated circuit
R1	39,000 Ω, 1/4 W, 5% resistor
C1	0.022 μF ceramic or metal film capacitor
C2	0.01 μF ceramic or metal film capacitor
C3	0.1 μF ceramic or metal film capacitor
B1	9V battery and holder
S1	SPST switch

A signal injector is used from the *back* to the *front* of an audio (or RF) circuit. For instance, to use the signal injector with an AM receiver apply the signal from the probe at the base of the output transistor. If that stage and everything after it operates correctly, the signal will be heard in the speaker. If the output stage proves to be OK, move back to the base of the driver transistor. The output signal will be higher if everything is working. Then apply the signal progressively toward the front of the circuit by injecting it at the volume control, detector stage, IF stages, and the mixer.

The circuit can be housed in a small plastic box and connected to the external circuits through a probe and an alligator clip.

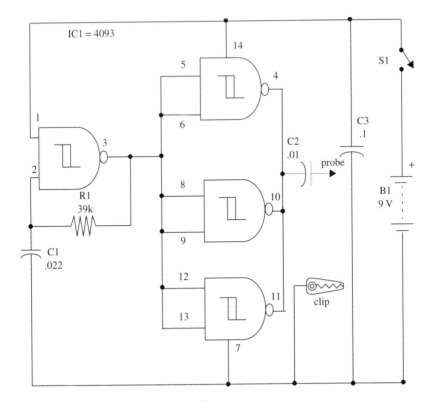

Figure 2.3 Signal Injector, schematic diagram.

Project 4: Touch-Controlled Oscillator I (E)

The pressure of your fingers on the sensor will determine the frequency of the tone produced by this oscillator. By changing your finger's pressure on the sensor, you can produce musical tones.

The circuit runs in frequencies between 100 and 1,000 Hz, depending on the sensor and C1 value. You can alter C1 in a wide range of values to get different sounds from the oscillator. Values between 470 pF and 0.1 μF can be used for experimentation.

The output transducer is a piezoelectric device, but you can use powerful stages as shown in other projects. The circuit should be powered with voltage sources between 6 and 12 V, and it drains only a few milliamperes with the recommended output transducer.

Figure 2.4 shows the complete circuit of the touch-controlled oscillator.

Parts List: Touch-Controlled Oscillator I

IC1	4093 CMOS integrated circuit
X1	Sensor (see text)
X2	Piezoelectric transducer or crystal earphone, Radio Shack 273-073 or equivalent
R1	1,000,000 Ω potentiometer
R2	10,000 Ω, 1/4 watt, 5% resistor
C1	1,200 pF ceramic capacitor
C2	100 μF, 12 WVDC electrolytic capacitor

Two small plates or screws form the touch sensor. If you use a conductive foam or two metal plates as sensor, change C1 to a large value (between 0.022 and 0.047 μF) to get a pressure-controlled oscillator.

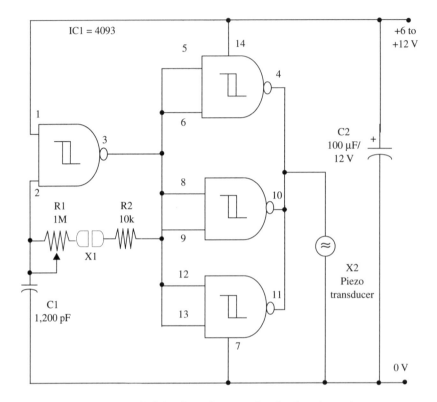

Figure 2.4   Touch-Controlled Oscillator I (sensor details given in text).

Project 5: Light-Controlled Oscillator I (E)

In this circuit, the output signal has a frequency that depends on the amount of light falling on the sensor, a light-dependent resistor (LDR). The frequency range, between 10 and 1,000 Hz, can be adjusted by R1 and also depends on C1 values. This capacitor can be altered in a wide range of values, from 0.01 to 0.1 µF.

The output transducer is a piezoelectric type, but other output stages can be used as transistor/loudspeaker stages. The circuit drains only a few milliamperes of current and can be powered from batteries.

Figure 2.5 shows the schematic diagram of this project.

Parts List: Light-Controlled Oscillator I

IC1	4093B CMOS integrated circuit
X1	Piezoelectric transducer or crystal earphone, Radio Shack 273-073 or equivalent
R1	100,000 Ω potentiometer
R2	Cadmium-sulfide photoresistor (LDR), Radio Shack 276-1657 or equivalent
C1	0.022 µF metal film or ceramic capacitor
C2	100 µF, 12 WVDC electrolytic capacitor

In this project, the photoresistor acts as a variable resistor controlling the generated frequency according the incident light. In the dark, the photoresistor resistance is very high, and the generated frequency is in the lower limit. With strong light, the photoresistor resistance is very low, and the generated frequency is in the upper limit.

For component values given in the schematic diagram, the frequency range is between 1 and 1,000 Hz. Capacitor values can be altered to change the frequency range. Values between 0.01 and 0.1 µF can be used for experimentation in this project.

By moving your hand in front of the photoresistor, you can alter the amount of incident light and produce musical tones.

Powerful output stages can be used, as in other projects shown in this book.

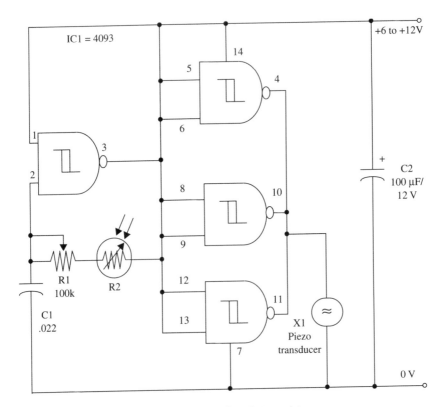

Figure 2.5   Light-Controlled Oscillator with a photo-resistor as sensor.

Project 6: Plasma Oscillator (E)

A flame is a conductive medium that can be used in a different feedback circuit to control the frequency of an audio oscillator. This configuration can be used to probe of the conductivity of a flame in science demonstrations or high school science projects.

The audio oscillator described in this project is controlled by the "fourth" state of the matter (*plasma*, or ionized gas) and can be used in various physics experiments.

The flame can be produced with a simple match, and the flickering effect can modulate the generated sound.

The circuit runs in frequencies between 1 Hz and 500 Hz, depending on the electrode construction and the flame position.

The complete schematic diagram is shown in Fig. 2.6.

Parts List: Plasma-Oscillator

IC1	4093 CMOS integrated circuit
X1	Plasma sensor (see text)
X2	Piezoelectric transducer or crystal earphone, Radio Shack 273-073 or equivalent
C1	1,200 pF ceramic capacitor
C2	100 µF, 12 WVDC electrolytic capacitor
S1	SPST switch
B1	6 to 12 V battery or power supply

Two wires, one placed near the other with about 1 inch of bare length, form the *plasma sensor.* The flame should touch the two bare wires to allow the feedback to operate the oscillator.

Powerful output stages can drive other transducers such as loudspeakers.

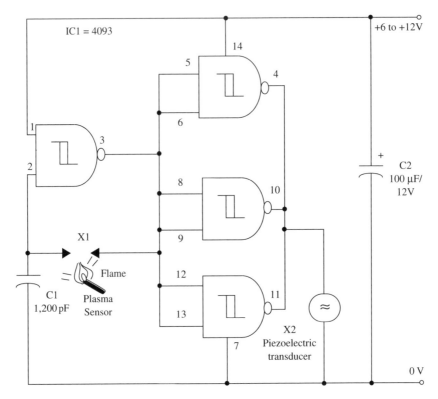

Figure 2.6 Current flows through the "plasma" produced by a flame. This is the feedback used in the oscillator.

Project 7: Insect Repellent (E) (P)

Some continuous sounds can repel insects (and other animals). The frequency and intensity depend on the application and can be determined experimentally.

The circuit described here generates a continuous sound that can be used to repel some types of insects. It can be powered by a 9 V battery or AA cells. The low current requirements will extend the life of the cells.

Figure 2.7 shows the schematic diagram of the insect repellent. A home-made printed-circuit board is shown in Fig. 2.8. All the components and the power supply can be housed in a small plastic box.

Parts List: Insect Repellent

IC1	4093 CMOS integrated circuit
X1	Crystal earphone or piezoelectric transducer, Radio Shack 273-073 or equivalent
R1	100,000 Ω, 1/4 W, 5% resistor
R2	10,000 Ω, 1/4 W, 5% resistor
C1	0.01 μF metal film or ceramic capacitor
C2	0.1 μF metal film or ceramic capacitor
S1	SPST switch
B1	6 V (four AA cells) or 9 V (battery)

Using the insect repellent is very easy. You only have to adjust the trimmer potentiometer R1 to produce a sound with a pitch that is appropriate to the insect you intend to repel. Experiment until you find the best sound to repel a specific insect. (If in doubt, ask the insect!)

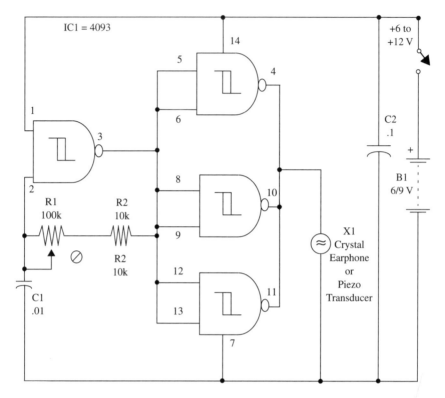

Figure 2.7 Schematic diagram of the Insect Repellent. The circuit can be powered by four AA cells or a 9 V battery.

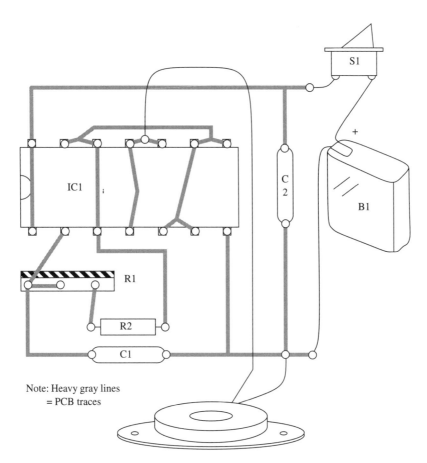

Figure 2.8   PCB layout for the Insect Repellent.

Project 8: Audio Generator (P)

This project produces a handy tool for audio (and RF) troubleshooting. The described audio generator produces squarewave signals between 100 and 1,000 Hz but can easily modified to operate in a wide range with some changes in the original circuit.

The output signal voltage can be adjusted from 0 V up to the power supply voltage, in a range up to 9 V. A complete diagram of the audio generator is shown in Fig. 2.9.

C1 and R1 determine the frequency range of the audio generator. C1 can be altered to other frequency ranges, but the best way to get a large-band audio generator is to connect a band switch with two or more capacitors as shown in Fig. 2.10.

Using capacitors with values suggested in this circuit, you can generate square signals up to 100 kHz.

All the components can be housed in a small plastic box as shown in Fig. 2.11. Frequency is adjusted by R1, and the amplitude is adjusted by R3.

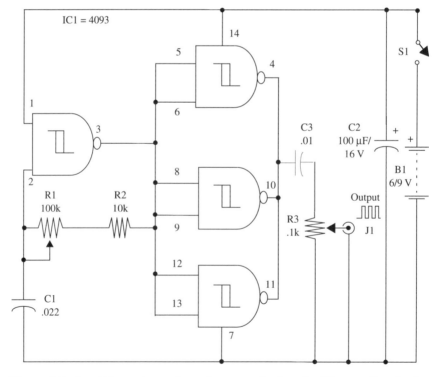

Figure 2.9 A switch can be used to select several values for C1, extending the frequency range.

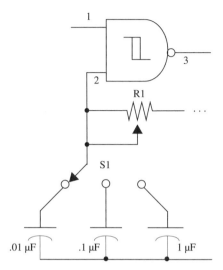

Figure 2.10 Band switch connection to the Audio Generator.

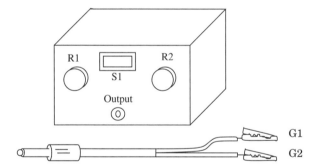

Figure 2.11 The Audio Generator is housed in a plastic box as shown above.

Parts List: Audio Generator

IC1	4093 CMOS integrated circuit
R1	100,000 Ω linear potentiometer
R2	10,000 Ω, 1/4 W, 5% resistor
R3	1,000 Ω linear potentiometer
C1	0.022 µF ceramic or metal film capacitor
C2	100 µF, 12 WVDC electrolytic capacitor
C3	0.01 µF ceramic or metal film capacitor
S1	SPST switch
B1	6 V (four AA cells) or 9 V (battery)

Project 9: Metronome I (E) (P)

This device produces an audio signal consisting of a series of "clicks." The clicks may be used as a metronome (their rate can be altered by R1) or as reference for gymnastic exercises or jogging.

The circuit can be powered from batteries and housed in a small plastic box, creating a portable unit.

The frequency range can be adjusted between 0.1 and 10 Hz but can easily be altered by changing C1. Values up to 2 µF can be used for large intervals between the produced pulses.

For the transducer, you can use a piezoelectric device (see parts list) or a crystal earphone. The transducers and IC drain a super-low current, extending the battery life. Figure 2.12 shows the schematic diagram of this device.

To calibrate the unit, synchronize your metronome with a commercial type or any known source, including a reference cassette recording.

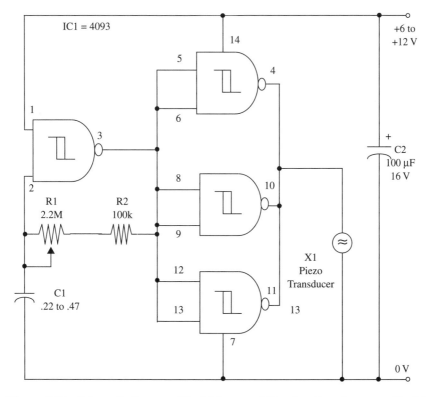

Figure 2.12   Schematic diagram of the Metronome I. The circuit can be powered by AA cells or a 9 V battery.

Parts List: Metronome I

IC1	4093 CMOS integrated circuit
X1	Piezoelectric transducer or crystal earphone, Radio Shack 273-073 or equivalent
R1	2,000,000 Ω, 1/4 W, 5% resistor
R2	100,000 Ω, 1/4 W, 5% resistor
C1	0.22 µF or 0.47 µF ceramic or metal film capacitor
C2	100 µF, 16 WVDC electrolytic capacitor

Project 10: Metronome II (E) (P)

This is a "high-power" version of the previous project, adding a simple transistorized output stage to the original circuit. With a Darlington output stage, the circuit can drive a loudspeaker with some hundreds milliwatts. You can also add a power FET to get a power output up to a few watts.

As the current drain is higher in this version, a power supply is recommended. Thus, it is best to use the unit as a fixed device supplied from the power mains.

As in Metronome I, the frequency range can be altered by changing C1. Capacitors up to 2 µF can be used, producing very low-frequency clicks.

The complete schematic diagram of the Metronome II is shown in Fig. 2.13.

Transistor Q1 must be mounted on a small heatsink. For best sound level the speaker should be installed into a loudspeaker enclosure.

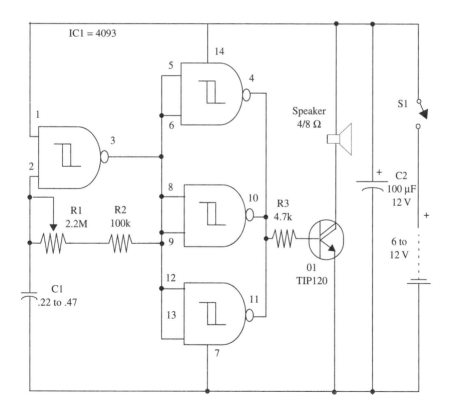

Figure 2.13 Metronome II. Transistor Q1 must be mounted on a heatsink.

This enclosure can also be used to house all the other components as shown in 2.15.

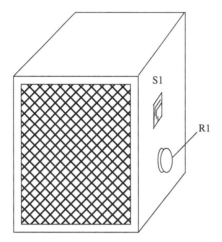

Figure 2.14 Loudspeaker and circuit are housed in the same enclosure, as shown above.

Parts List: Metronome II

IC1	4093 CMOS integrated circuit
Q1	TIP120 or equivalent Darlington power transistor (Or power-FET)
SPKR	4/8 Ω, 2- to 4-inch loudspeaker
S1	SPST switch
B1	6 V to 12 V power supply
R1	2,200,000 Ω, 1/4 W, 5% resistor
R2	100,000 Ω, 1/4 W, 5% resistor
R3	4,700 Ω, 1/4 W, 5% resistor
C1	0.22 μF or 0.47 μF ceramic or metal film capacitor
C2	100 μF, 12 WVDC electrolytic capacitor

Project 11: Ultrasonic Generator I (E) (P)

Dogs, mice, rats, bats, some birds, and other animals can hear sounds with frequencies up to 40,000 Hz. The circuit we propose here produces continuous ultrasound in frequencies above the human limit, in a range between 18,000 Hz and 40,000 Hz. The device can be used to scare dogs and other animals in biological experiments and many other applications.

The recommended piezoelectric transducer has its maximum output power in frequencies between 700 and 3,000 Hz, but it will also operate in higher frequencies, emitting less power.

The recommended power supply is formed by four AA cells or a 9 V battery. The very low current drain extends the power supply life.

Our project runs at approximately 18,000 to 40,000 Hz, but you can easily alter this range changing C1 or R1. C1 can be altered in the range between 470 pF and 0.001 µF, and R1 can be altered up to 100,000 Ω. The 4093 IC will oscillate in frequencies up to 500 kHz.

A complete circuit diagram of the ultrasonic transmitter is shown in Fig. 2.15.

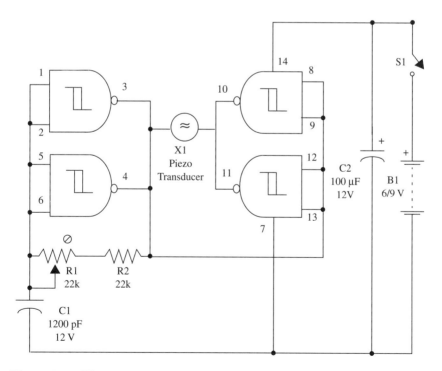

Figure 2.15 Ultrasonic Generator I. This circuit will produce signals in the 18 to 40 kHz range.

Parts List: Ultrasonic Generator I

IC1	4093 CMOS integrated circuit
X1	Piezoelectric transducer or crystal earphone, Radio Shack 173-073 or equivalent
R1	22,000 Ω pot/trimmer
R2	22,000 Ω, 1/4 W,5% resistor
C1	1,200 pF metal film or ceramic capacitor
C2	100 µF, 12 WVDC electrolytic capacitor
S1	SPST switch
B1	6 V (four AA cells) or 9 V (battery)

The circuit can be housed in a small plastic box. The transducer will be fixed in the front panel.

Project 12: Ultrasonic Generator II (E) (P)

Two 4093 ICs can be used to perform a powerful ultrasonic generator, as we show in this project. The circuit drives a piezoelectric transducer or crystal earphone with some tens of milliwatts and operates in a frequency range between 18,000 and 40,000 Hz. The frequency range can also be altered by changing the value of C2. The upper limit of the circuit is 1 MHz.

Biological experiments related to animal behavior and conditioning can be conducted using this oscillator. The power supply is formed by four AA cells or a 9 V battery. The circuit drains only few milliamperes, extending the power supply life to several weeks.

R1 can be connected in series with a 47,000 Ω potentiometer to allow frequency adjustment in a wide range.

The schematic diagram or the Ultrasonic Generator II is shown in Fig. 2.16.

For the transducer, you can also use a piezoelectric tweeter. This component has a small output transformer inside, as shown in Fig. 2.17. You have to disconnect the transformer to use the tweeter in this project.

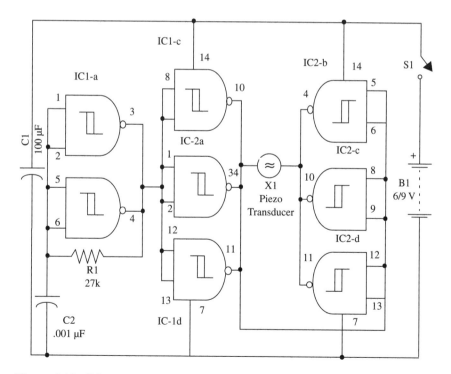

Figure 2.16 Schematic diagram of Ultrasonic Generator II.

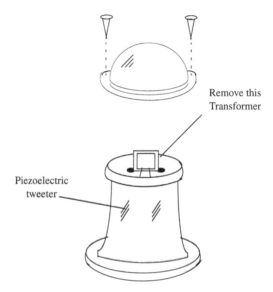

Figure 2.17   The small transformer should be removed for this project.

Parts List: Ultrasonic Generator II

IC1, IC2	4093 CMOS integrated circuit
X1	Piezoelectric Transducer or crystal earphone, Radio Shack 273-073 or equivalent
R1	27,000 Ω, 1/4 W, 5% resistor
C1	100 μF, 12 WVDC electrolytic capacitor
C2	0.001 μF ceramic or metal film capacitor
S1	SPST Toggle or momentary switch
B1	6 V (four AA cells) or 9 V (battery)

Project 13: Ultrasonic Generator III (E) (P)

This third version of the ultrasonic generator uses a piezoelectric tweeter and a transistorized output stage to give a powerful output signal. The transducer, driven by a two-transistor output stage, can produce about 400 mW of ultrasonic signal.

The circuit is powered by four AA cells or a 9 V battery, and its current drain is about 50 mA.

The frequency can be set by R1 in a range between 18,000 and 40,000 Hz. You can alter this frequency range by changing C1. Values between 470 and 4,700 pF can be used experimentally.

Although the optimum performance for tweeters, as the output devices in this project, is between 10,000 Hz and 20,000 Hz, these transducers, as we experimentally proved, can also operate with a fair output up to 40,000 Hz.

It is not necessary to disconnect the small internal transformer in this application as we did in the previous project. You can also use a small ultrasonic transducer with impedance ranging from 4 to 100 Ω.

A schematic diagram of Ultrasonic Generator III is shown in Fig. 2.18. The device can be housed in a small plastic box, as shown in Fig. 2.19.

Parts List: Ultrasonic Generator III

IC1	4093 CMOS integrated circuit
Q1	2N2222 general-purpose NPN silicon transistor (Radio Shack 276-1617)
Q2	2N2907 general-purpose PNP silicon transistor
X1	4/8 Ω piezoelectric tweeter, Radio Shack 40-1383
S1	SPST momentary or toggle switch
B1	6 V (four AA cells) or 9 V (battery)
R1	47,000 Ω pot/trimmer
R2	10,000 Ω, 1/4 W, 5% resistor
R3	2,200 Ω, 1/4 W, 5% resistor
C1	1,200 pF ceramic capacitor
C2, C3	100 µF, 12 WVDC electrolytic capacitor

The reader should keep in mind that piezoelectric tweeters have a directional emission characteristic. Frequency adjustment can be made using a frequency meter connected to IC pin 4. A square wave is produced, and its amplitude is the power supply voltage.

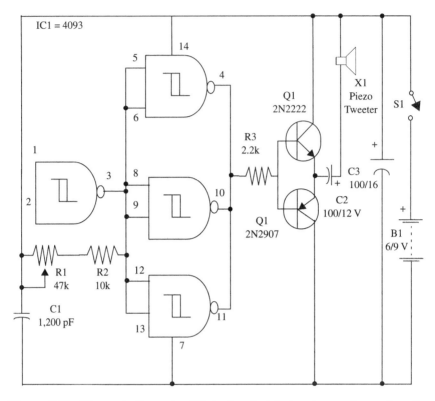

Figure 2.18 Ultrasonic Generator III. A piezoelectric tweeter can be used as the transducer.

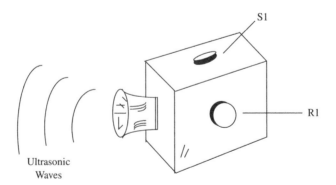

Figure 2.19 The project can be housed in a plastic box, as shown above.

Project 14: High-Power Ultrasonic Generator (E) (P)

This circuit can produce an ultrasonic output of a few watts with a piezoelectric or other type of tweeter. The operating frequency ranges from 18,000 to 40,000 Hz but can also be altered by changing C1. Higher values for C1 will give sound in the audible range, which allows the unit to be used in alarms and other applications. In this case, the tweeter can be replaced by a common loudspeaker.

The circuit drains some hundreds of milliamperes from a 9 V to 12 V power supply. Batteries are recommended for momentary operation only.

The reader can use this device to scare dogs and other animals by installing the unit near waste deposits or other places where unwanted animals can be present.

Ultrasonic operation is achieved with a C1 value between 470 and 2,200 pF. Audible operation will occur with C1 in the range between 0.01 and 0.022 µF.

A schematic diagram of the powerful ultrasonic generator is shown in Fig. 2.20.

The transistors must be mounted on heatsinks. All components can be housed in a plastic box. The value of capacitor C1 depends on the frequency range you want to generate.

Parts List: Powerful Ultrasonic Generator

IC1	4093 CMOS integrated circuit
Q1, Q3	TIP31 NPN silicon power transistor
Q2, Q4	TIP32 PNP silicon power transistor
SPKR	4/8 Ω tweeter or loudspeaker (see text)
R1	100,000 Ω potentiometer
R2	10,000 Ω, 1/4 W, 5% resistor
R3, R4	2,200 Ω, 1/4 W, 5% resistor
C1	1,200 pF (ultrasonic) or 0.022 µF (sonic) metal film or ceramic capacitor
C2	100 µF, 12 WVDC electrolytic capacitor

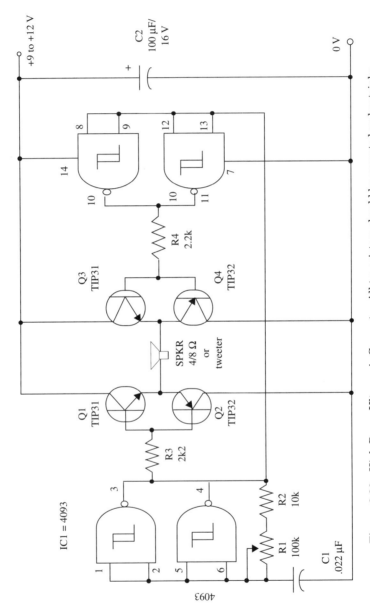

Figure 2.20 High-Power Ultrasonic Generator. All transistors should be mounted on heatsinks.

Project 15: Match Oscillator (E)

A home-made 400 MΩ resistor is used in this interesting experimental oscillator. The match resistor has some unusual characteristics and can be used to detect humidity. Depending on the absorbed atmospheric humidity, the match resistor changes its resistance and thus the oscillator frequency. Several scientific experiments and demonstrations in science fairs can be devised using this circuit.

Building Rx is easy: wind two or three turns of a 4 or 5 cm long piece of bare wire on each extremity of a wooden match, leaving a length of 2 or 3 cm of the wire to be used as terminals. Press the wire to affix it without the need of glue or other media. The resistor is ready to be used.

The described oscillator will produce "clicks" in a transducer. The click rate is determined by C1 and Rx values. Using a frequency meter, you can use the circuit to measure atmospheric humidity.

Only two of the four 4093 gates are used in this project, and the output transducer is a piezoelectric transducer. You can alter the output stage using transistors to drive small loudspeakers.

A complete schematic diagram of this experiment is shown in Fig. 2.21. The click rate depends on the absorbed humidity. You can conduct experiments with this sensor using water vapor, heaters, and so on.

Parts List: Match Oscillator

IC1	4093 CMOS integrated circuit
X1	Piezoelectric transducer or crystal earphone, Radio Shack 273-073 or equivalent
Rx	Match resistor (see text)
C1	120 pF ceramic capacitor
C2	100 μF, 12 WVDC electrolytic capacitor

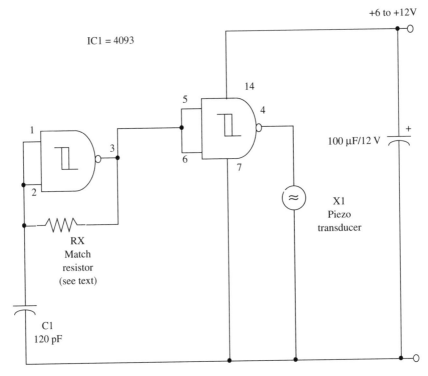

Figure 2.21 Match Oscillator. Rx is a 500 MΩ resistor built with a match.

Project 16: Morse Code Tone Generator (P)

If you're an aspiring ham radio operator or Boy Scout working toward an award in signaling or radio, here is a useful circuit that is very easy to get working. The circuit is inexpensive, and a simple sending key is shown in Fig. 2.22. It comprises a strip of springy brass mounted on a wooden block, and two wood screws.

The brass is bent upward into a crank shape, and a plastic button is glued to the free end with epoxy adhesive to form a knob. Beneath the free end of the spring is a large brass pin or a round-headed brass screw. Key contacts are made to this screw, with one end of the screw securing the other end of the brass strip.

The tone is adjusted by means of R1, and volume by R3. The frequency of the free-running oscillator can be adjusted in a wide band, but output tones between 500 and 1,000 Hz are the most pleasant and the least tiring if you intend to practice for extended periods. The complete schematic circuit of this device is shown in Fig. 2.23.

You can power this device with power supplies ranging from 6 V to 12 V, but with the 12 V power supply, the output transistor must be mounted on a heatsink.

All the components can be housed in a wooden box for better sound level, and portability.

A simple version can use a piezoelectric transducer directly connected to outputs b/c and d of the IC.

Parts List: Morse Tone Generator

IC1	4093 CMOS integrated circuit
Q1	TIP31 NPN silicon power transistor
SPKR	4/8 Ω, 2-inch loudspeaker
M1	Morse key (see text)
B1	6 V (four AA cells) or 9 V (battery)
R1	100,000 Ω, 1/4 W, 5% resistor
R2	22,000 Ω, 1/4 W, 5% resistor
R3	50 to 100 Ω rheostat or wire wound potentiometer
R4	4,700 Ω, 1/4 W, 5% resistor
C1	0.022 µF metal film or ceramic capacitor
C2	100 µF, 16 WVDC electrolytic capacitor

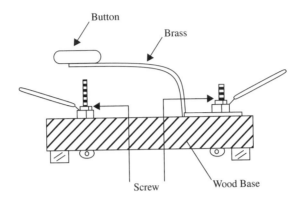

Figure 2.22 Morse key construction.

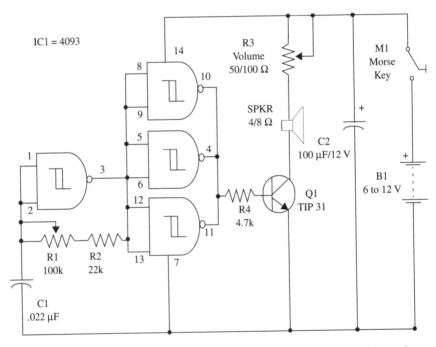

Figure 2.23 Morse Code Tone Generator. Transistor Q1 must be mounted on a heat-sink.

Project 17: 100 Hz to 1,000 Hz Oscillator (E)

This is a two-gate oscillator that produces a square signal in the range between 100 and 1,000 Hz. The device can be used as a signal injector or as the basis for several projects. Alarms, insect repellents, ultrasonic generators are some examples of projects that can use this basic configuration.

The frequency range can be altered by changing C1. Ultrasonic signals will be produced with capacitor in the 470 pF to 2,200 pF range, and ultra-low frequency signals will be produced with capacitors in the range between 0.22 and 0.47 µF.

A piezoelectric transducer is used in this project, but it can be altered with an additional transistorized driver stage. This stage will drive a loudspeaker, producing a more powerful audio signal.

The circuit can be powered with sources in the range between 5 and 12 V. Only few milliamperes are drained by the circuit, allowing battery operation.

A schematic diagram of the device is shown in Fig. 2.24. Frequency adjustments can be made by connecting a frequency counter to pin 10 or 11 (IC1).

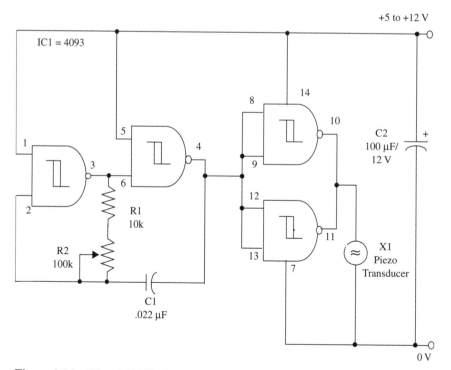

Figure 2.24   100 to 1,000 Hz Oscillator using a two-gate design.

Parts List: 100 Hz to 1,000 Hz Generator

IC1	4093 CMOS integrated circuit
X1	Piezoelectric transducer or crystal earphone, Radio Shack 273-073 or equivalent
R1	10,000 Ω, 1/4 W, 5% resistor
R2	100,000 Ω potentiometer
C1	0.022 µF metal film or ceramic capacitor
C2	100 µF, 12 WVDC electrolytic capacitor

Project 18: Touch-Controlled Oscillator II (E)

The two-gate oscillator shown in this project has its frequency controlled by your finger's pressure on a dual metal-plate sensor. The circuit will oscillate in a frequency range from 10 Hz to 2,000 Hz, depending on the components used.

Of course, this frequency range can be altered easily by changing C1. We suggest that the reader experiment with values between 0.01 and 0.47 µF.

The recommended output transducer is an earphone or a piezoelectric transducer, but this can also be altered by using a powerful transistorized output stage to drive loudspeakers or tweeters.

Required power supply voltage is in the range between 5 and 12 V. As the current drain is very low, you can also use battery to supply the unit.

A schematic diagram of Touch-Controlled Oscillator II is shown in Fig. 2.25.

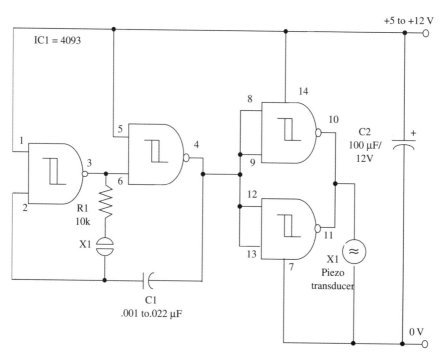

Figure 2.25   This touch-controlled oscillator produces tones with the pitch depending on finger pressure on X1.

Parts List: Touch-Controlled Oscillator II

IC1	4093B CMOS integrated circuit
X1	Piezoelectric transducer or crystal earphone, Radio Shack 273-073 or equivalent
R1	10,000 Ω, 1/4 W, 5% resistor
R2	100,000 Ω potentiometer
C1	0.022 μF metal film or ceramic capacitor
C2	100 μF, 12 WVDC electrolytic capacitor

Project 19: Light-Controlled Oscillator II (E)

The frequency of the signal produced by this oscillator depends on the amount of light that falls onto a sensor. For the sensor, we used a light-dependent resistor (LDR) or photoresistor, as it is also called. The transistorized output stage directly drives a low-impedance loudspeaker. If you want, you can eliminate that stage and connect a piezoelectric transducer to IC pins 10 and 11.

The circuit can be powered with voltages between 5 and 12 V. If using 9 V to 12 V supplies, the transistor must be mounted on a heatsink.

The frequency range depends on C1 and the amount of light on the sensor. You can alter the frequency range by changing C1. Values in the range between 0.01 and 0.022 µF will produce sounds in the audio range. If you want to produce "clicks," with a rate depending on the amount of light on the sensor, you could experiment with capacitors in the range between 0.22 and 1 µF. Metal film or ceramic capacitors can be used in this experiment.

You can produce musical tones by moving your hand in front of the sensor, controlling the amount of light on it. Another interesting application is a light-frequency converter, sending information corresponding to light intensity to a remote frequency counter. A correspondence table with values of frequency and amount of light should be used for the remote reader.

A schematic diagram of the device is shown in Fig. 2.26.

Parts List: Light-Controlled Oscillator II

IC1	4093 CMOS integrated circuit
LDR	Photoresistor LDR, Radio Shack 276-1657 or equivalent
SPKR	4/8 Ω loudspeaker
Q1	TIP31 or 2N2218 silicon medium power transistor (see text)
R1	10,000 Ω, 1/4 W, 5% resistor
R2	4,700 Ω, 1/4 W, 5% resistor
C1	0.01 to 0.022 µF metal film or ceramic capacitor
C2	100 µF, 12 WVDC electrolytic capacitor

Transistor Q1 depends on the power supply voltage. If you're using voltages between 6 and 9 V, the recommended transistor is the 2N2218 with a small heatsink. If you're using voltages supplies between 9 and 12 V, the recommended transistor is the TIP31 with a heatsink.

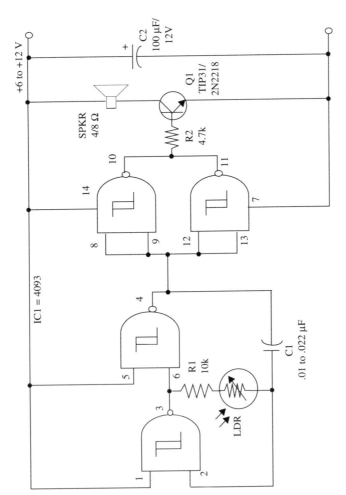

Figure 2.26 Another sound-to-light converter using a two-gate oscillator.

Project 20: 100 kHz to 1 MHz CW Transmitter (E)

This RC oscillator's upper frequency limit is about 1 MHz. The circuit we describe will operate as a short-range experimental transmitter in the range between 100 kHz and 1 MHz.

The reader can use the device as a standard for receiver calibration or for troubleshooting RF stages as a simple signal generator. As the output waveform is square, harmonics extend the useful upper frequency limit up to 100 MHz. Even FM receiver can receive the signal produced by this simple circuit.

With a small antenna connected to the transmitter, you can receive the signals at distances up to 3 ft (1 m). Thus, you don't need to use wires or physical connections on the receivers you're calibrating.

Frequency adjustment is made by R1 in the AM range (LF and MF). A Morse key can be connected between the circuit and power supply to create a CW transmitter.

A complete schematic diagram of the transmitter is shown in Fig. 2.27. The antenna is a length of wire of between 1 and 6 ft.

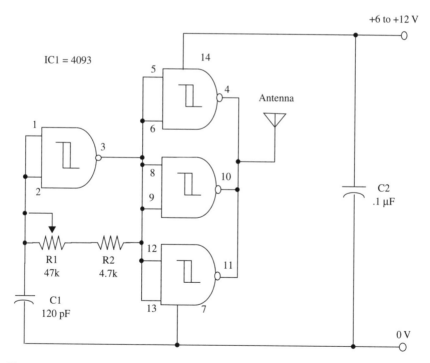

Figure 2.27   This circuit is a 100 kHz to 1 MHz transmitter. Notice that no coils are used.

Parts List: 100 kHz to 1 MHz CW Transmitter

IC1	4093 CMOS integrated circuit
R1	47,000 Ω potentiometer
R2	4,700 Ω, 1/4 W, 5% resistor
C1	120 pF ceramic capacitor
C2	0.1 µF ceramic or metal film capacitor

To use the device, place an AM receiver tuned to an unoccupied point of the spectrum, near the transmitter. Adjust R1 in the transmitter while pressing the Morse key until you detect the signal.

Project 21: Modulated 100 kHz to 1 MHz Transmitter (E)

This project uses two oscillators to produce a modulated output in the range between 100 kHz and 1 MHz. The device can be used as an experimental telegraphic transmitter or as a modulated signal generator for troubleshooting RF stages in AM, FM, and other receivers.

The circuit will oscillate in frequencies up to 1 MHz, and modulation is fixed within about 1 kHz. You can alter the modulation range by adding a potentiometer and a series resistor in place of R1. A 100,000 Ω potentiometer and a 10,000 Ω resistor produce modulation signals in the range between 100 and 1,000 Hz. Capacitor C1 can also be altered to change this modulation frequency.

The circuit can be powered by batteries, as the drained current is very low (only a few milliamperes), extending the power supply life.

As a coil-less circuit, the frequency adjustment is made by a potentiometer. The antenna is a wire 1 to 6 feet in length.

A schematic diagram of the transmitter is shown in Fig. 2.28.

Parts List: Modulated 100 kHz to 1 MHz Transmitter

IC1	4093B CMOS integrated circuit
K1	Morse key (see text)
R1	39,000 Ω, 1/4 W, 5% resistor
R2	47,000 Ω potentiometer
R3	4,700 Ω, 1/4 W, 5% resistor
C1	0.022 µF ceramic or metal film capacitor
C2	120 pF ceramic capacitor
C3	0.1 µF ceramic or metal film capacitor

For the Morse key, the reader can use the same device as described in Project 16. The operating frequency is adjusted by R2. See project 20 to more information on applications of this device.

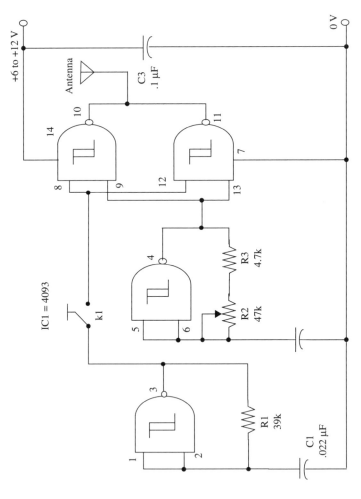

Figure 2.28   This coil-less transmitter will send a modulated signal in the range of 100 kHz to 1 MHz.

Project 22: 3 to 4 MHz CW Transmitter (E)

An LC-oscillator can run in frequencies up to 4 MHz using only one of 4093 IC's gates. The circuit we show here can be used as an experimental low-power transmitter operating in the 80-meter amateur band or as a signal generator.

The circuit drains only a few milliamperes from the power supply, which can be formed using a 9 V battery or 6 AA cells.

As in the CMOS oscillators, the upper frequency limit depends on the power supply voltage, in this application a minimum of 9 V is required for a 3 to 4 MHz operation. With a 6 V supply, the frequency upper limit will fall to 2 MHz.

Experimental applications employ a single antenna wire 1 to 6 feet long, but you can also connect an appropriate external antenna to operate in the 80-meter ham band.

A complete schematic diagram of the transmitter is shown in Fig. 2.29.

Parts List: 3 to 4 MHz CW Transmitter

IC1	4093 CMOS integrated circuit
L1	47 µH RF choke (see text)
C1	4.7 or 10 pF ceramic capacitor
C2	6 to 50 pF trimmer, Radio Shack 272-1340 or equivalent
C3	0.1 µF ceramic capacitor
S1	SPST or Morse key
B1	9 to 12 V power supply

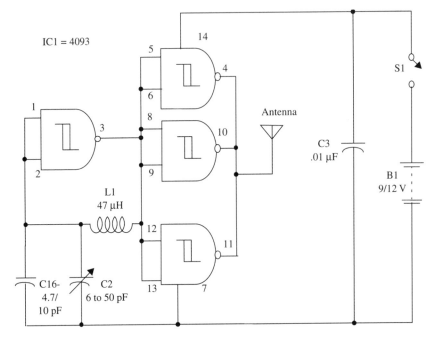

Figure 2.29   Experimental transmitter for frequencies up to 3 or 4 MHz.

Project 23: Tone-Modulated 4 to 4 MHz Transmitter (E)

One of the 4093 IC's gates is used in this transmitter to create an audio oscillator that modulates the transmitting signal. Other gate is used as a RF oscillator operating between 3 and 4 MHz. The third and fourth gates are used to form a buffer that drives the antenna with a tone-modulated 3 to 4 MHz signal.

The high-frequency oscillator is controlled by C1, and the audio-frequency oscillator is controlled by R1.

Using a short antenna (1 to 6 ft), the signals can be received in distances up to 10 ft.

Figure 2.30 shows the complete schematic diagram of the tone-modulated transmitter.

Parts List: Tone-Modulated 3 to 4 MHz Transmitter

IC1	4093 CMOS integrated circuit
L1	47 µH micro-choke
C1	10 pF ceramic capacitor
C2	6 to 50 pF trimmer, Radio Shack 272-1340 or equivalent
C3	0.022 µF ceramic or metal film capacitor
C4	0.1 µF ceramic or metal film capacitor
R1	100,000 Ω potentiometer
R2	10,000 Ω, 1/4 W, 5% resistor

The Morse key is connected between the power supply and the positive rail of the transmitter. L1 can also be a home-made coil. It is formed with 40 turns of 24 or 26 gauge insulated wire on a ferrite rod, with the coil having a 1/8-inch I.D.

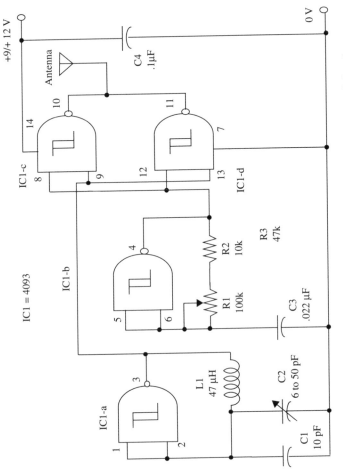

Figure 2.30   This 80-m band experimental transmitter is tone modulated.

Project 24: Beeper I (E) (F)

You can assemble this simple device for very little cash and can use it in several interesting applications. You can use your beeper in alarms, toys, games, and as a low-power warning enunciator for your electronic equipment.

The circuit will produce an intermittent sound in a frequency of about 1 Hz. A piezoelectric transducer is used but, in the next version (Beeper II), we will use a powerful output stage to drive loudspeakers.

Audio tone is generated by IC1-a, and its frequency is determined by R1 and C1. You can easily alter this frequency by changing R1. Values of R1 between 22 and 100 kΩ can be used experimentally or, if you prefer, a 100 kΩ potentiometer in series with a 10 kΩ resistor can be used.

The repetition rate is governed by R2 and C2 and can also be altered by changing C2. Values between 0.15 and 0.47 µF, and even 1 µF, can be investigated. It is important to test for the ideal value for this component, which depends on the intended application.

A schematic diagram of Beeper I is given in Fig. 2.31.

Parts List: Beeper I

IC1	4093 CMOS integrated circuit
X1	Piezoelectric transducer or crystal earphone, Radio Shack 273-072 or equivalent
R1	39,000 Ω, 1/4 W, 5% resistor
R2	2,200,000 Ω, 1/4 W, 5% resistor
C1	0.022 µF metal film or ceramic capacitor
C2	0.47 to 1 µF metal film or ceramic capacitor
C3	100 µF, 12 WVDC electrolytic capacitor

The power supply current drain is very low (only a few milliamperes), so you can use AA cells or a 9 V battery to supply the unit. You can also connect the unit to the power supply of the equipment to which you will add it.

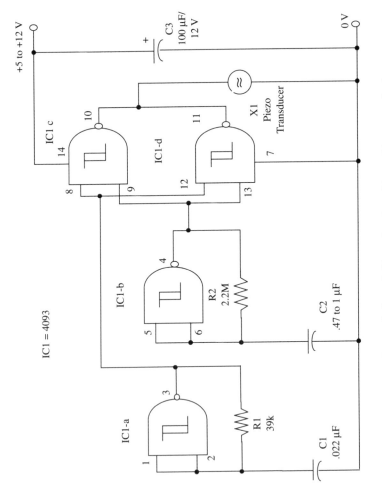

Figure 2.31   Beeper I. C1 and C2 can be altered in a wide range of values.

Project 25: Power Beeper II (E) (P)

This is a powerful version of the Beeper I using a transistorized output stage to drive small loudspeakers. Some hundreds of milliwatts are produced by this unit, which can be used as an alarm, in toys, and for several other applications. The repetition rate is adjusted by R1, and the audio tone is adjusted by R3.

Both C1 and C2 can be altered to change the tone and rate characteristics of the project. You can experiment with several values according the intended application.

The complete schematic diagram of the unit is shown in Fig. 2.32.

Parts List: Power Beeper II

IC1	4093 CMOS integrated circuit
Q1	2N2222 general purpose NPN silicon transistor
Q2	2N2907 general purpose PNP silicon transistor
SPKR	4/8 Ω loudspeaker
S1	SPST switch
B1	6 V (four AA cells) or 9 V (battery)
R1	2,200,000 Ω potentiometer
R2	100,000 Ω, 1/4 W, 5% resistor
R3	100,000 Ω potentiometer
R4	10,000 Ω, 1/4 W, 5% resistor
R5	4,700 Ω, 1/4 W, 5% resistor
C1	1 μF to 2.2 μF metal film capacitor
C2	0.022 μF metal film or ceramic capacitor
C3	47 μF, 12 WVDC electrolytic capacitor
C4	100 μF, 12 WVDC electrolytic capacitor

The current drain is about 50 mA. For better audio, the loudspeaker should be installed in a small speaker enclosure.

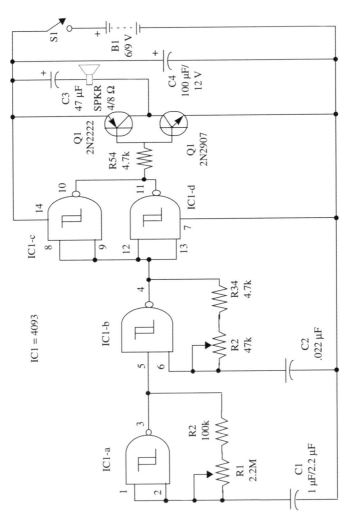

Figure 2.32   Beeper II. Rate and tone are controlled by R1 and R3.

Project 26: Beeper III (E) (P)

This one-tone beeper has a different output stage that drives a piezo-electric transducer or crystal earpiece. It works as the previous project but, due the complementary output stage, it will provide a little more power than Project 24, Beeper I.

The frequency of the two oscillators can also be altered by changing capacitors C1 and C2. The circuit can be powered with voltages in the range between 6 V and 9 V. Batteries can be used due the low current drain.

A schematic diagram of Beeper III is shown in Fig. 2.33.

Parts List: Beeper III

IC1	4093 CMOS integrated circuit
X1	Piezoelectric transducer or crystal earphone, Radio Shack 273-073 or equivalent
S1	SPST switch
B1	6 V (four AA cells) or 9 V (battery)
R1	2,200,000 Ω, 1/4 W, 5% resistor
R2	47,000 Ω, 1/4 W, 5% resistor
C1	0.22 μF to 0.47 μF ceramic or metal film capacitor
C2	0.022 μF ceramic or metal film capacitor
C3	100 μF, 12 WVDC electrolytic capacitor

Large values of C1 will provide a slow repetition rate, and large values of C2 will produce lower-frequency tones.

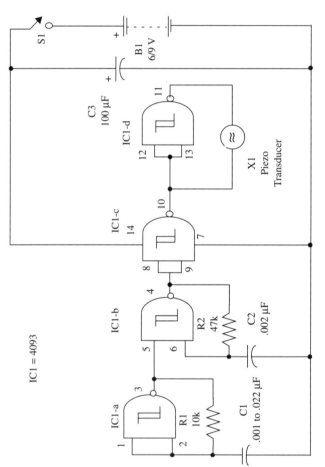

Figure 2.33 Beeper III. A piezoelectric transducer is used in this project.

Project 27: Two-Tone Beeper I (E) (P)

This circuit will generate a two-tone warning at a rate of about 1 Hz and can be used as an experimental siren, as part of an alarm, or in toys, games, and other applications.

The tones have frequencies of about 1 kHz and 1.8 kHz, but you can easily alter them by changing C2 and C3. C1 determines the alternation rate and can also be varied in a large range of values.

The circuit is powered with four AA cells or a 9 V battery, but it can also function with other voltages in the range between 5 and 12 V. Only a few milliamperes are drained from the power supply, extending the life of the batteries (if employed).

The complete schematic diagram of the unit is shown in Fig. 2.34.

Parts List: Two-Tone Beeper I

IC1	4093 CMOS integrated circuit
X1	Piezoelectric transducer or crystal earphone, Radio Shack 273-073 or equivalent
S1	SPST switch
B1	6 V (four AA cells) or 9 V (battery)
R1	2,200,000 Ω, 1/4 W, 5% resistor
R2	47,000 Ω, 1/4 W, 5% resistor
R3	27,000 Ω, 1/4 W, 5% resistor
C1	0.47 μF ceramic or metal film capacitor
C2	0.022 μF ceramic or metal film capacitor
C3	0.022 μF ceramic or metal film capacitor
C4	100 μF, 12 WVDC electrolytic capacitor

R1, R2, and R3 can be replaced by a potentiometer in series with resistors. R1 is replaced by a 2.2 MΩ potentiometer and a 100 kΩ series resistor. R2 and R3 are replaced by 100 kΩ potentiometers and 10 kΩ series resistors. This way, the device can be used as a simple sound synthesizer.

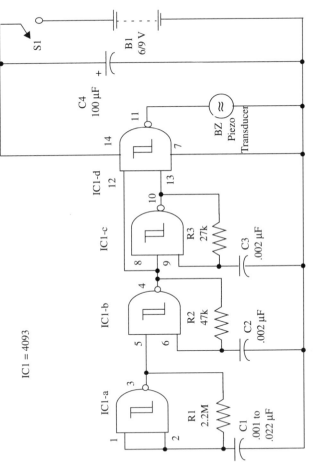

Figure 2.34   Two-Tone Beeper I. This unit drives a crystal earphone or piezoelectric transducer.

Project 28: Power Beeper IV (E) (P)

This circuit produces an intermittent one-tone beep in a loudspeaker. Power is between 500 mW and 2 W, depending on the power supply voltage and loudspeaker impedance. This circuit can be used in alarms and similar applications.

Current drain is about 500 mA, precluding the use of small batteries or AA cells as power supplies. The tone is adjusted by R3, and the repetition rate by R1.

A schematic diagram of the unit is shown in Fig. 2.35.

Parts List: High-Power Beeper IV

IC1	4093 CMOS integrated circuit
Q1	TIP120 NPN Darlington power transistor
SPKR	4/8 Ω, 4-inch loudspeaker
R1	100,000 Ω potentiometer
R2	10,000 Ω, 1/4 W, 5% resistor
R3	2,200,000 Ω, 1/4 W, 5% resistor
R4	100,000 Ω, 1/4 W, 5% resistor
C1	0.022 μF ceramic or metal film capacitor
C2	0.22 or 0.47 μF metal film capacitor
C3	100 μF, 12 WVDC electrolytic capacitor

Transistor Q1 must be mounted on a heatsink. To get more power from this circuit, you can replace Q1 with a power FET. Any type with current rated to 2 A or more can be used in this project.

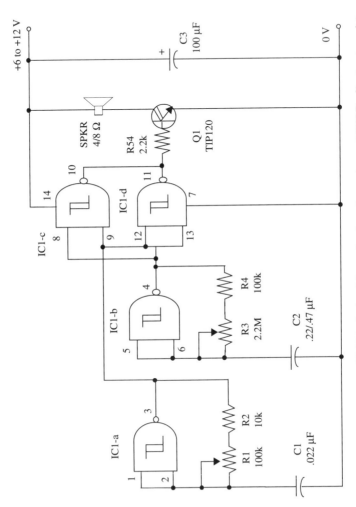

Figure 2.35   Power Beeper IV. This device drives a loudspeaker using an NPN Darlington transistor.

Project 29: Complementary Beeper V (E) (P)

A powerful version of the previous beepers can be made using a complementary transistor output stage. Output power up to 4 W can be obtained from this circuit, which can be used in alarms, warning systems, games, and so on.

The circuit doesn't have frequency adjustments, but it easily can be altered to incorporate tone and rate controls. R1 and R2 can be replaced by potentiometers in series with resistors. R1 is replaced by a 100 kΩ potentiometer and a 10 kΩ series resistor, and R2 is replaced by a 2.2 MΩ or a 4.7 MΩ potentiometer with a 100 kΩ series resistor.

The transistors used depend on the power supply voltage. In the 6 V version, you can use the 2N2222 and 2N2907 output pair, general purpose silicon transistors. With 9 V (or higher) supplies, you have to use the pair formed by TIP31 and TIP32 silicon power transistors, which should be mounted on heatsinks.

For better sound reproduction, the loudspeaker should be installed in a enclosure. Speakers of 4-inch diameter and more will give better results in this circuit.

Current drain depends on the supply voltage and can range from 200 mA to 1A. A schematic diagram of the device is shown in Fig. 2.36.

Parts List: Complementary Beeper V

IC1	4093 CMOS integrated circuit
Q1	2N2222 or TIP31 NPN transistor (see text)
Q2	2N2907 or TIP32 PNP transistor (see text)
SPKR	4/8 Ω, 4-inch loudspeaker
R1	47,000 Ω, 1/4 W, 5% resistor
R2	2,200,000 Ω, 1/4 W, 5% resistor
R3	4,700 Ω, 1/4 W, 5% resistor
C1	0.022 μF ceramic or metal film capacitor
C2	0.47 μF or 1 μF metal film capacitor
C3	220 μF, 16 WVDC electrolytic capacitor
C4	100 μF, 16 WVDC electrolytic capacitor

The position of the polarized components (e.g., C3 and C4) should be observed.

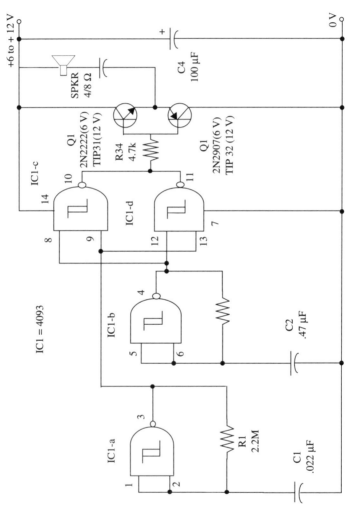

Figure 2.36 Complementary Beeper V. TIP31/TIP32, if used, should be mounted on heatsinks.

Project 30: Two-Tone Siren (E) (P)

This circuit produces a two-tone, high-level output to a loudspeaker. The two-tone frequencies and alternation rate can be adjusted by potentiometers, and the device can be supplied by a 12 V car battery or power supply.

You can use this circuit in cars, alarms, and many other applications. The current drain is up to 1 A, requiring a heavy-duty supply to power the unit.

Output transistors must be mounted on heatsinks, and fuse F1 protects the unit against shorts when used in cars.

A schematic diagram of the Two-Tone Siren is shown in Fig. 2.37.

Parts List: Two-Tone Siren

IC1	4093 CMOS integrated circuit
Q1	TIP31 NPN silicon power transistor
Q2	TIP32 PNP silicon power transistor
SPKR	4/8 Ω, 4-inch loudspeaker
F1	2-A fuse
R1	2,200,000 Ω potentiometer
R2	100,000 Ω, 1/4 W, 5% resistor
R3, R5	100,000 Ω, 1/4 W, 5% resistors
R4, R6	10,000 Ω, 1/4 W,5% resistors
R7	4,700 Ω, 1/4 W, 5% resistor
C1	0.47 μF ceramic or metal film capacitor
C2, C3	0.022 μF ceramic or metal film capacitors
C4	220 μF, 16 WVDC electrolytic capacitor
C5	100 μF, 16 WVDC electrolytic capacitor

The speaker should be installed in an enclosure for better sound.

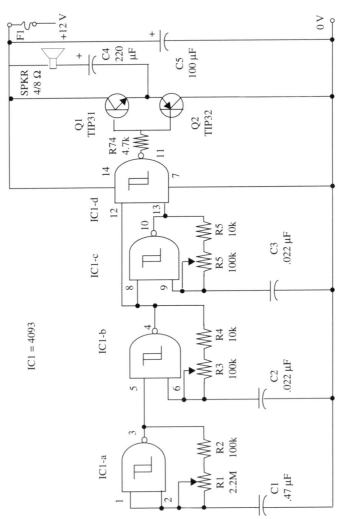

Figure 2.37 Two-Tone Siren. Transistors Q1 and Q2 should be mounted on heatsinks.

Project 31: Frequency-Modulated Siren (E) (P)

Acting on a capacitor charging, one oscillator can modulate another oscillator, changing its frequency as shown in this project. The tone thus runs from high to low and vice versa in a rate determined by the first oscillator frequency.

The circuit can be used as a siren, as part of alarms or games, and in many other applications. With a 12 V power supply, the output is up to a few watts, representing an excellent audio level in a loudspeaker.

The basic circuit we show has fixed tone, modulation rate, and depth, but you can alter these characteristics by changing some components. To alter the modulation rate, you have to change R1 or C1; to alter tone, you have to change R4 or C3. Modulation depth can be altered by changing R2 and also R3.

If you want to conduct some interesting experiments in sound generation, you can replace all of the indicated resistors by potentiometers in series with other resistors. The rule to make this replacement is simple: the potentiometer should have 2× or the same value as the substituted resistor, and the series resistor should have about 1/10 of the potentiometer value.

For instance, if you replace a 100 kΩ resistor with a 100 or 220 kΩ potentiometer, the series resistor should be a 10 or 22 kΩ unit.

The output transistors depend on the power supply voltage. For a 6 V power supply, you can use the 2N2222/2N2907 pair, both general purpose silicon transistors. But if you intend to use a 9 V or a 12 V power supply, the transistors should be power units such as the TIP31/TIP32 pair, and they must be mounted on heatsinks.

Current drain depends on the power supply voltage. It typically ranges from 100 mA to 500 mA.

A schematic diagram of the Frequency-Modulated Siren is shown in Fig. 2.38. The loudspeaker should be installed into an enclosure for better sound reproduction.

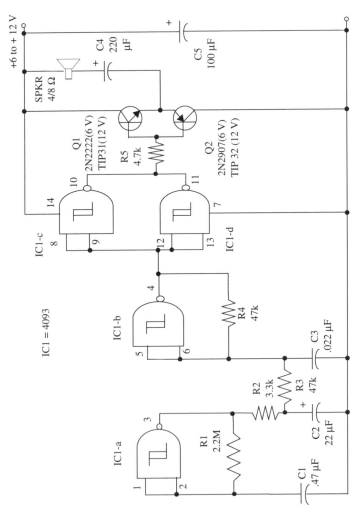

Figure 2.38 FM Siren. Output transistors should be mounted on heatsinks in the 12 V version.

Parts List: PARTS LIST Frequency-Modulated Siren

IC1	4093B CMOS integrated circuit
Q1	2N2222 or TIP31 (see text) NPN transistor
Q2	2N2907 or TIP32 (see text) PNP transistor
SPKR	4/8 Ω, 4-inch loudspeaker
R1	2,200,000 Ω, 1/4 W, 5% resistor
R2	3,300 Ω, 1/4 W, 5% resistor
R3	47,000 Ω, 1/4 W, 5% resistor
R4	47,000 Ω, 1/4 W, 5% resistor
R5	4,700 Ω, 1/4 W, 5% resistor
C1	0.47 µF ceramic or metal film capacitor
C2	22 µF, 12 WVDC electrolytic capacitor
C3	0.022 µF ceramic or metal film capacitor
C4	220 µF, 16 WVDC electrolytic capacitor
C5	100 µF, 16 WVDC electrolytic capacitor

Project 32: Complex-Sound Generator (E) (P)

Combining four different squarewave signals, you can produce a complex output such as produced by this device. The circuit can be used to experiment new sounds or for troubleshooting or evaluation of audio equipment.

Both frequency and individual oscillator level are controlled by potentiometers in a large range of values. Frequencies can range from 50 to 1,000 Hz with the components listed, and the signal level can be adjusted from 0 V to the power supply voltage.

Connect the output of the generator to the oscilloscope input and see how many waveforms you can generate by varying the eight potentiometers' positions. You will see that, if you adjust the first potentiometer to $2 \times f$ (2f), the third to $3 \times f$ (3f), the fourth to $4 \times f$ (4f), and so on, and adjust the output level potentiometers in decreasing steps, you can synthesize a sinewave with good precision. (See Fourier for more details!) See Fig. 2.39.

If you want to alter the frequency range of this device, you only have to change the capacitors. The circuit can produce signals up to 1 MHz. The minimum recommended value for C1 to C3 is 120 pF.

A complete schematic diagram of the Complex-Sound Generator is shown in Fig. 2.40.

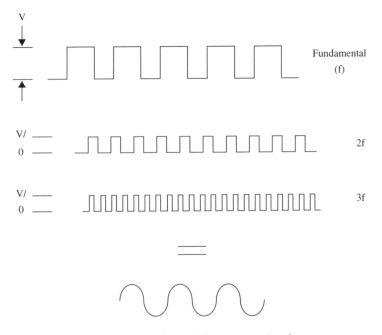

Figure 2.39 A sinewave can be synthesized from square signals.

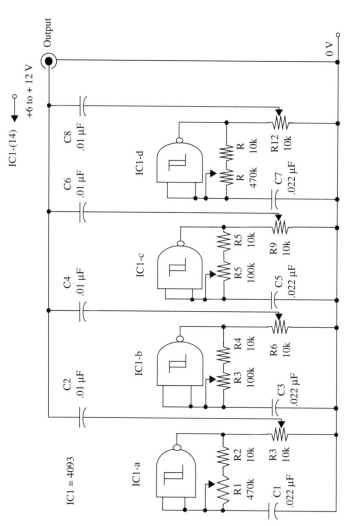

Figure 2.40   Complex-Sound Generator. This circuit produces complex signals synthesized from squarewaves.

Parts List: Complex-Sound Generator

IC1	4093 CMOS integrated circuit
R1, R4, R7, R10	470,000 Ω potentiometers
R2, R5, R8, R11	10,000 Ω, 1/4 W, 5% resistor
R3, R6, R9, R12	10,000 Ω potentiometers
C1, C3, C5, C7	0.022 µF ceramic or metal film capacitors
C2, C4, C6, C8	0.01 µF ceramic or metal film capacitors

The output signals carried through an output jack.

Project 33: Variable-Interval Beeper (E)

All of the beeper projects we have seen in this book so far have had a 50 percent duty cycle. That means that the duration of the produced tone is the same as the interval between tones for all frequencies. At low frequencies, this characteristic produces long beeps that are not ideal for some applications.

With some improvements, the 4093 oscillator can be altered to produce constant-duration pulses in any frequency range, and the frequency range can be adjusted by a simple potentiometer.

The circuit we show here provides an example of how to do that. Our beeper produces short (or long) pulses in a rate from about 3 or 4 per second to one per 3 or 4 seconds. The tone also can be adjusted in a wide range of frequencies from 100 to 1,000 Hz.

The operation of the circuit is easy to explain. IC1-a is connected as an inverter and will oscillate as described in the introductory part of this book. When power is on, C1 charges through R3, and thus the high output of this gate triggers on the second oscillator (IC1-b).

During C1 charge, a tone is produced. When V_p is reached in C3, IC1-a output goes low, and then the IC1-b oscillator stops. C1 begins to discharge via R1, R2, and D1 until V_n is reached. Then, the IC1-a output goes high again, and a new cycle begins.

Note that C1 charges through D2 and R3 and discharges through R1, R2, and D2. Thus, as R1 controls the interval between the output high pulses, R3 determines their duration (see Fig. 2.41).

Potentiometer R4 adjusts tone frequency, and the output transducer is a piezoelectric device or a crystal earphone. R3 can be altered for shorter or longer pulses in the range between 100 kΩ and 1 MΩ (audio pulses).

A schematic diagram of the beeper is given in Fig. 2.42. By replacing X1 with a 10,000 Ω potentiometer, you can convert this beeper into a variable duty cycle audio generator.

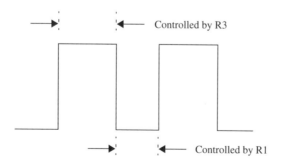

Figure 2.41 R1 and R2 control the duty cycle for the generated signal.

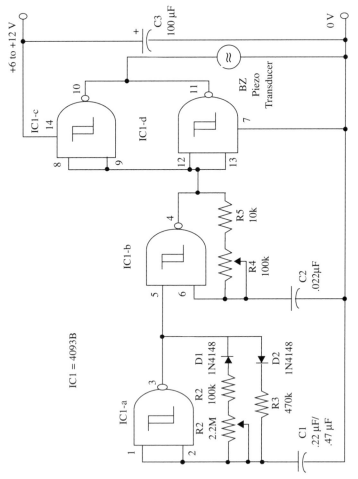

Figure 2.42 Complex-Sound Generator. This circuit produces complex signals synthesized from squarewaves.

Parts List: Variable-Interval Beeper

IC1	4093 CMOS integrated circuit
X1	Piezoelectric transducer or crystal earphone, Radio Shack 273-073 or equivalent
D1, D2	1N4148 general purpose silicon diodes
R1	2,200,000 Ω potentiometer
R2	100,000 Ω, 1/4 W, 5% resistor
R3	470,000 Ω, 1/4 W, 5% resistor
R4	100,000 Ω potentiometer
C1	0.22 μF or 0.47 μF ceramic or metal film capacitor
C2	0.022 μF ceramic or metal film capacitor
C3	100 μF, 16 WVDC electrolytic capacitor

Project 34: Variable Duty Cycle Siren (E) (P)

Regulated intervals between audio tones are produced by this siren. Thus, the audio tone frequency and length is adjusted by potentiometers.

Audio tone duration is adjusted by R2, and the separation between audio pulses is adjusted by R3. The audio tone frequency is adjusted by R5.

The project has a transistorized output stage to drive a loudspeaker with the power depending on the power supply voltage. Outputs up to 4 W can be obtained with a 12 V power supply. Current drain in this case is up to 1 A. Circuit operation is as explained in previously in Project 33.

The output transistors employed will depend on the power supply voltage. For a 5 to 6 V supply voltage, you can use the 2N2222/2N2907 pair of general purpose silicon transistors. But if you're using a 9 to 12 V supply, you should use the TIP31/TIP32 pair of power silicon transistors, mounted on heatsinks. Figure 2.43 shows the schematic diagram of this project.

Parts List: Variable Duty Cycle Siren

IC1	4093 CMOS integrated circuit
Q1	2N2222 or TIP31 NPN transistor (see text)
Q2	2N2907 or TIP32 PNP transistor (see text)
SPKR	4/8 Ω, 4-inch loudspeaker
D1, D2	1N4148 general purpose silicon diodes
R1, R3	2,200,000 Ω, 1/4 W, 5% resistors
R2, R4	100,000 Ω, 1/4 W, 5% resistors
R5	100,000 Ω potentiometer
R6	10,000 Ω, 1/4 W, 5% resistor
R7	4,700 Ω, 1/4 W, 5% resistor
C1	0.22 µF or 0.47 µF ceramic or metal film capacitor
C2	0.022 µF ceramic or metal film capacitor
C3	220 µF, 16 µF WVDC electrolytic capacitor
C4	100 µF, 16 WVDC electrolytic capacitor

The duty cycle can be adjusted in a range between 5 and 95%.

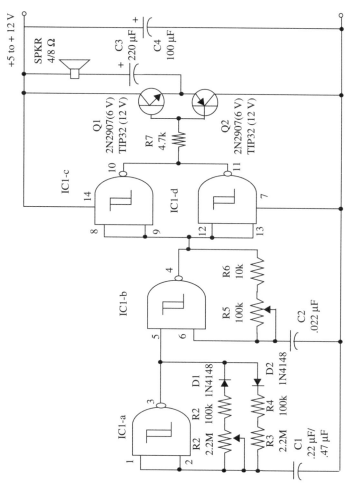

Figure 2.43   Variable Duty Cycle Siren. Output transistors depend on power supply voltage.

Project 35: Touch-Triggered Siren (E)

Some touch-controlled projects we have seen in this book are skin resistance-regulated oscillators rather than skin resistance-triggered, as in this case. The device produces an intermittent sound when the sensor is touched. Tone frequency and separation between tone pulses are fixed. The audio tone has a 1,000 Hz frequency and is repeated at a rate of one per second.

Circuit operation is easy to explain: when the sensor is touched, pin 2 (IC1-a) goes low and the output goes high, triggering the two oscillators, formed by IC1-b and IC1-c, to the "on" state.

Modulation and tone signals produced by these two oscillators are combined in the fourth gate (IC1-d) and applied to the output transducer. As in previous projects you can replace the piezoelectric transducer with transistorized output stages to drive loudspeakers.

The sensor is made with two small metal plates that should be touched at the same time to trigger the circuit. A schematic diagram of the device is shown in Fig. 2.44.

This device can be used as a touch alarm or in toys to produce sound effects when touched.

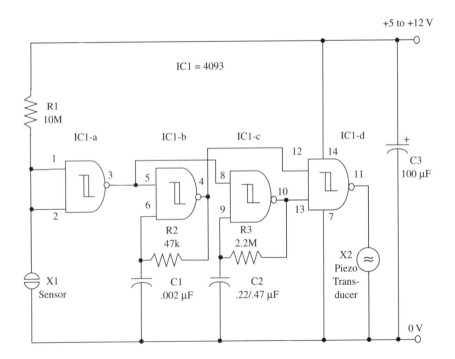

Figure 2.44 This siren triggers "on" when the sensor is touched.

Parts List: Touch-Triggered Siren

IC1	4093 CMOS integrated circuit
X1	Metal plate sensor (see text)
X2	Piezoelectric transducer or crystal earphone, Radio Shack 273-073 or equivalent
R1	10,000,000 Ω, 1/4 W, 5% resistor
R2	47,000 Ω, 1/4 W, 5% resistor
R3	2,200,000 Ω, 1/4 W, 5% resistor
C1	0.022 µF metal film or ceramic capacitor
C2	0.22 µF or 0.47 µF metal film or ceramic capacitor
C3	100 µF, 12 WVDC electrolytic capacitor

Project 36: Sound Machine (E) (P)

Would you like to create different "space" sound effects with a simple device? Sirens, birds, space monsters, and guns can be imitated with the project we describe here.

The combined positions of the four potentiometers and two switches will give you unlimited possibilities for sound creation.

Our circuit is composed of two oscillators: one running at a low frequency between 0.2 and 2 Hz and used as a modulator. The other oscillator runs in the audio frequency range between 100 and 1,000 Hz and is used for tone generation.

The potentiometers are used to control the frequency of the oscillators and also modulation depth. The switches can connect the oscillator in four different ways:

1. In the first position, IC1-b, the audio oscillator is in the free-running mode. Thus, we have a continuous sound, adjusted in frequency by R5. In this connection, S1 is in A, and S2 is in D.

2. In the second mode (S1 in A and S2 in C), the slow oscillator frequency modulates IC1-b, the audio-frequency oscillator. Modulation is adjusted by R1, and depth is adjusted by R3 and R4. The device will generate a continuous sound with variations in frequency.

3. In the third condition, with S1 in B and S2 in A, the device will produce an interrupted sound with a rate given by R1 and a tone by R5. R3 and R4 don't affect the circuit in this condition.

4. Finally, in the fourth mode, S1 in B and S2 in C, we have an interrupted tone that is modulated in frequency. The interruption rate is given by R1, tone frequency by R5, and modulation depth by R3 and R4.

The capacitors can be altered if you want to create new effects. The capacitors C1, C2, and C3 can be altered within a wide range of values. Create new sound experiences! A schematic diagram of the sound-machine is shown in Fig. 2.45.

The choice of which transistors to use depends on the power supply voltage. If the power supply is between 5 and 6 V, you should use the 2N2222/2N2907 pair of general purpose silicon transistors. But if you're using a 9 to 12 V power supply, you should use the pair TIP31/TIP32 silicon power transistor. These power transistor should be mounted on heatsinks.

The loudspeaker should be mounted in an enclosure for better sound reproduction.

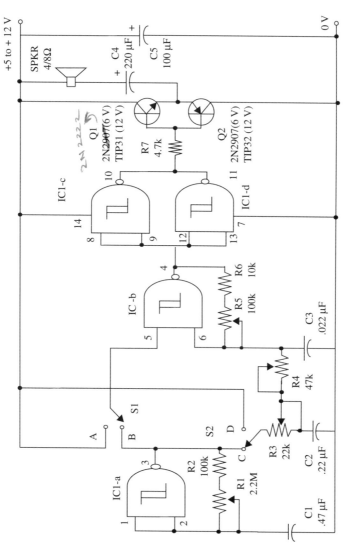

Figure 2.45   Sound Machine. You can create "space" sounds using this circuit.

Parts List: Sound Machine

IC1	4093 CMOS integrated circuit
Q1	2N2222 or TIP31 NPN silicon transistor (see text)
Q2	2N2907 or TIP32 PNP silicon transistor (see text)
SPKR	4/8 Ω, 4-inch loudspeaker
S1, S2	SPDT switches
R1	2,200,000 Ω potentiometer
R2	100,000 Ω, 1/4 W, 5% resistor
R3	22,000 Ω potentiometer
R4	47,000 Ω potentiometer
R5	100,000 Ω potentiometer
R6	10,000 Ω, 1/4 W, 5% resistor
C1	0.47 µF ceramic or metal film capacitor
C2	22 µF, 12 WVDC electrolytic capacitor
C3	0.022 µF ceramic or metal film capacitor
C4	220 µF, 16 WVDC electrolytic capacitor
C5	100 µF, 16 WVDC electrolytic capacitor

3

Lamp and LED Projects

This chapter provides light-effect projects involving signaling, test, and other special circuits using lamps and LEDs. Some of the projects also include sound effects. The reader should keep in mind that all of this book's projects can be combined in different ways to create other devices.

As in the previous chapter, we show only basic projects. More complex ones are shown in the final chapter, combining sound, light, digital, and timing circuits. Many will use components shown in previous chapters.

As in Chapter 1, all of these projects are easy to build and also are intended either to educate you about components and circuits [designated with an **(E)**] or to provide a practical device that can be used in your home, car, scientific endeavors, and so on [designated with a **(P)**].

Project 37: LED Flasher I (E)

A simple LED flasher can be built with a 4093 used as a low-frequency oscillator and driving a LED. IC1-a runs in a low frequency, determined by R1 adjustment, in the range between 0.1 and 5 Hz. This oscillator drives the other three 4093 gates where the LED is connected.

R3 depends on the power-supply voltage. With power supplies between 5 and 6 V, R3 is a 470 Ω resistor. With power supplies ranging from 9 to 12 V, R1 is a 1,000 Ω resistor.

Figure 3.1 shows the schematic diagram of the LED flasher.

Take special care with positioning of polarized components such as the LED and capacitor C2.

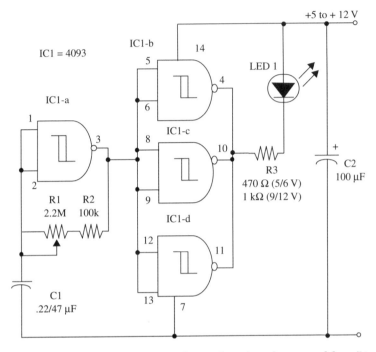

Figure 3.1 This LED flasher has a 50% duty cycle and can be powered from 5 to 12 V power supplies.

Parts List: LED Flasher I

IC1	4093 CMOS integrated circuit
Led1	Common red, yellow, or green LED
R1	2,200,000 Ω potentiometer
R2	100,000 Ω, 1/4 W, 5% resistor
R3	470 Ω, 1/4 W, 5% resistor (5 to 6 V power-supplies)
R4	1,000 Ω, 1/4 W, 5% resistor (9 to 12 V power supplies)
C1	0.22 μF or 0.47 μF ceramic or metal film capacitor
C2	100 μF, 16 WVDC electrolytic capacitor

Project 38: 4093 IC Tester (E)

This circuit will determine you if your 4093 IC is good or defective. It is a simple test circuit consisting of a low-frequency oscillator and a three-LED driver.

If the oscillator section is not good (IC1-a), the LEDs will not flash. But if the oscillator section is good and one of the other gates not, the corresponding LED will not flash but the other LEDs will flash at a rate of about 1 Hz. A schematic diagram of the 4093 IC test is shown in Fig. 3.2.

Parts List: 4093 IC Tester

IC1	4093 CMOS integrated circuit
LED1, LED2, LED3	Common red, green, or yellow LEDs
R1	2,200,000 Ω, 1/4 W, 5% resistor
R2, R3, R4	1,200 Ω, 1/4 W, 5% resistors
C1	0.22 µF ceramic or metal film capacitor
C2	100 µF, 16 WVDC electrolytic capacitor

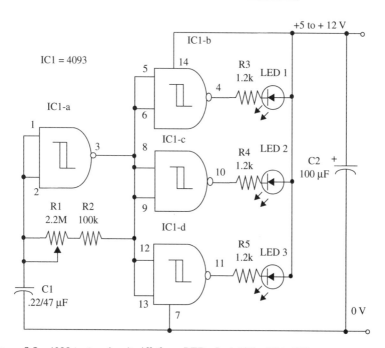

Figure 3.2 4093 tester circuit. All three LEDs flash if the IC is OK.

The circuit is mounted on a solderless board as shown in Fig. 3.3. Proper positioning of polarized components, such as the LEDs and capacitor C2, must be observed.

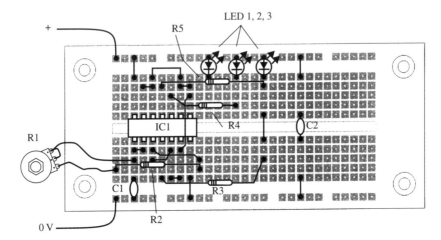

Figure 3.3 Component placement on a solderless board.

Project 39: 6/12 V Incandescent Lamp Flasher (E) (P)

This device will flash small 6 or 12 V lamps in a rate adjustable between 0.1 and 2 Hz by a potentiometer. A Darlington power transistor will drive lamps with current rates up to 1 A. The device can be used as a component of alarms or in cars, trailers, and so on. It is powered by 6 or 12 V batteries.

C1 can be altered to change the frequency range. High values will give lower flash rates. Values up to 2.2 μF can be used for experimentation. A schematic diagram of the device is shown in Fig. 3.4.

Transistor Q1 must be mounted on a heatsink. The lamp should be mounted in a suitable base. For experimental purposes, we suggest the E-10 base (Radio Shack 272-357 or equivalent). The lamps can be a #46 (250 mA, 6.3 V, Radio Shack 272-1130) or a #147 (Radio Shack 272-1134, 14 V, 200 mA).

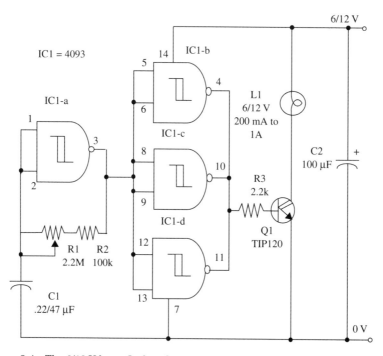

Figure 3.4   The 6/12 V lamp flasher. Q1 must be mounted on a heatsink.

Parts List: 6/12 V Incandescent Lamp Flasher

IC1	4093 CMOS integrated circuit
L1	6/12 V, 200 mA to 1 A incandescent lamp (see text)
Q1	TIP120 NPN Darlington transistor
R1	2,200,000 Ω, 1/4 W, 5% resistor
R2	100,000 Ω, 1/4 W, 5% resistor
R3	2,200 Ω, 1/4 W, 5% resistor
C1	0.22 or 0.47 µF ceramic or metal film capacitor
C2	100 µF, 16 WVDC electrolytic capacitor

Project 40: Variable Duty Cycle Incandescent Lamp Flasher (E) (P)

Duty cycles of 50 percent are given by the circuit in Project 39 and others appearing in this book. But there are some applications where you need different duty cycles—e.g., for reduction of energy consumption.

An adjustable duty cycle in the range between 5 and 95 percent can be obtained from this circuit. Also, the frequency range is adjustable between 0.1 and 5 Hz. C1 can be altered to change frequency range, and R2 and R4 can be altered to change duty cycle range.

The circuit can be powered from supplies ranging from 6 to 12 V, and the output power is sufficient to drive incandescent lamps up to 1 A.

This circuit works as follows: C1 charges through R3 and R4 and discharges through R1 and R2. Thus, R1 controls the discharge or length of the low output pulse, and R2 controls the length of the high output pulse. IC1-b, c, and d are used to drive the power output stage.

A schematic diagram of the lamp flasher is shown in Fig. 3.5.

Transistor Q1 must be mounted on a heatsink. Polarized components must be in their proper positions.

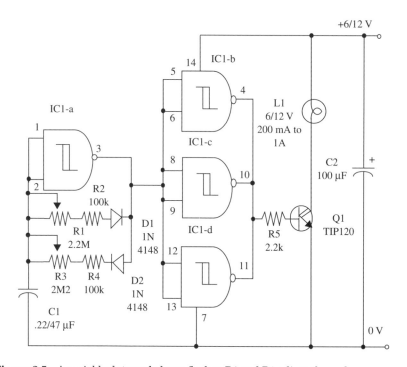

Figure 3.5   A variable duty cycle lamp flasher. R1 and R1 adjust the cycle.

Parts List: Variable Duty Cycle Lamp Flasher

IC1	4093 CMOS integrated circuit
Q1	TIP120 NPN Darlington power transistor
L1	200 mA to 1 A, 6 V or 12 V incandescent lamp (see text)
D1, D2	1N4148 general purpose silicon diodes
R1, R3	2,200,000 Ω potentiometers
R2, R4	100,000 Ω, 1/4 W, 5% resistor
R5	2,200 Ω, 1/4 W, 5% resistor
C1	0.22 µF or 0.47 µF ceramic or metal film capacitor
C2	100 µF, 16 WVDC electrolytic capacitor

The lamp must be mounted in a base. For experimental purposes, the E-10 base is suitable (Radio Shack 272-357 or equivalent). In a more complex signaling system, several lamps can be wired in parallel.

Project 41: Flasher with Beeper (E)

This sound-and-light warning device uses both a piezoelectric transducer to produce audio tones (in a rate adjusted by the user) and light flashes from an LED, also adjustable by the user.

Beeps and flashes are controlled by potentiometer R1 in a rate between 0.1 and 5 Hz, and tone frequencies are adjusted by potentiometer R3 in a range between 100 and 1,000 Hz.

For a powerful sound output, you can use one of the many output stages suggested in previous projects. Also, you can drive powerful lamps with power output transistors. A Darlington transistor such as the TIP115 can drive incandescent lamps up to 1 A.

The device can be used as a component of alarms or games, in cars, and for various other applications. Of course, the experimental version will teach you about this kind of circuit.

A schematic diagram of the Flasher with Beeper is shown in Fig. 3.6.

The tone and flash rate can be altered by changing the values of C1 and C2. Perform experiments according what you intend to do with the device whether it is used in applications other than the suggested ones.

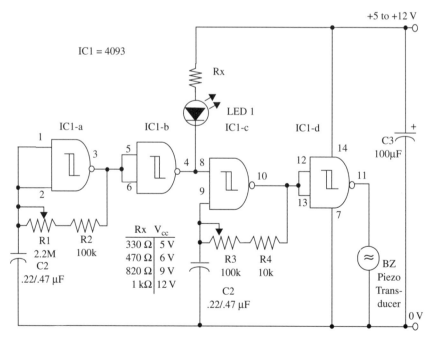

Figure 3.6 Beeps and flashes are produced by this circuit.

Parts List: Flasher With Beeper

IC1	4093 CMOS integrated circuit
Led1	Red, green, or yellow common LED
X1	Piezoelectric transducer or crystal earphone, Radio Shack 273-073 or equivalent
R1	2,200,000 Ω potentiometer
R2	100,000 Ω, 1/4 W, 5% resistor
R3	100,000 Ω potentiometer
R4	10,000 Ω, 1/4 W, 5% resistor
C1	0.22 μF or 0.47 μF ceramic or metal film capacitor
C2	0.022 μF ceramic or metal film capacitor
C3	100 μF, 16 WVDC electrolytic capacitor

Project 42: LED Flasher II (E)

This circuit uses a two-gate oscillator to produce 50 percent duty cycle flashes in a common LED. The circuit can be powered by 5 to 12 V power supplies and used in several practical applications. You can use it as part of alarms, warning systems, games, toys, and so on.

Rx depends on the power-supply voltage according table in the schematic diagram. For more output power, you can use a transistorized output stage as shown in other projects. The powerful output transistor stage using a TIP120 can drive lamps up to 1 A.

A schematic diagram of Led Flasher II is given in Fig. 3.7.

Parts List: Led Flasher II

IC1	4093 CMOS integrated circuit
LED1	Red, green, or yellow common LED
R1	2,200,000 Ω or 3,300,000 Ω, 1/4 W, 5% resistor (see text)
Rx	1/4 W, 5% resistor, according the power-supply voltage (see text)
C1	0.22 µF or 0.47 µF ceramic or metal film capacitor
C2	100 µF, 16 WVDC electrolytic capacitor

The flash rate can be altered by changing C1 and R1. R1 could have values between 1 and 10 MΩ, and C2 between 0.22 and 2.2 µF. Perform experiments to find the value that will give the best performance.

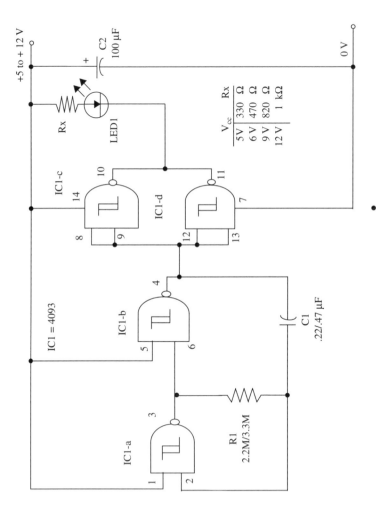

Figure 3.7 Beeps and flashes are produced by this circuit.

Project 43: Dual Variable Duty Cycle Flasher (E)

This circuit uses a variable duty cycle, two-gate oscillator to drive a dual-color LED or two different LEDs at a rate between 0.1 Hz and 5 Hz.

The power supply can range from 5 to 12 V, and the "on" time of each LED can be adjusted independently to a wide range of values.

This device can be used as component of alarms; in warning systems, toys, games; or as an experiment to teach about astable multivibrators.

In this application, two gates perform as an astable multivibrator. The astable frequency is given by C1, C2, R1, R2, and R3. Both, R1 and R3 act on the produced signal and on the "on" time of each gate output.

C1 and C2 can be altered in a wide range of values according the intended application for the device.

A schematic diagram of the device is shown in Fig. 3.8.

Parts List: Dual Variable Duty Cycle Flasher

IC1	4093 CMOS integrated circuit
LED1, LED2	Dual-color or red and green common LEDs, Radio Shack 276-012 or equivalent dual-color part
R1, R3	1,000,000 Ω potentiometers
R2, R4	10,000 Ω, 1/4 W, 5% resistors
R5, R6	1/4 W, 5% resistors (see text) for values according power-supply voltage
C1, C2	10 µF to 470 µF, 16 WVDC electrolytic capacitors (see text)
C3	100 µF, 16 WVDC electrolytic capacitor

Proper positioning of the polarized components (electrolytic capacitors and LEDs) should be observed when mounting.

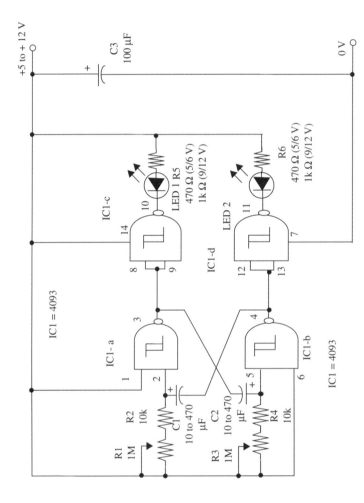

Figure 3.8 LED Flasher II. This circuit drives a single LED.

Project 44: Dual-LED Flasher I (E)

This circuit has a 50 percent duty cycle and can drive a bicolor LED or two common LEDs. The LEDs will flash alternately at a rate that can be adjusted between 0.1 and 5 Hz.

The LEDs are driven by two transistor, thus you can wire up to five LEDs with the corresponding resistor to each transistor. The flash rate is adjusted by R1, and you can alter C1 to change the frequency range.

The circuit can also drive small incandescent lamps with currents up to 100 mA, without any modification in the given basic project. You only have to observe the lamp voltage and replace the LED and the series resistor with it.

A schematic diagram of the Dual-LED Flasher is shown in Fig. 3.9. Proper positioning of the polarized components (LEDs, electrolytics, and transistors) must be observed.

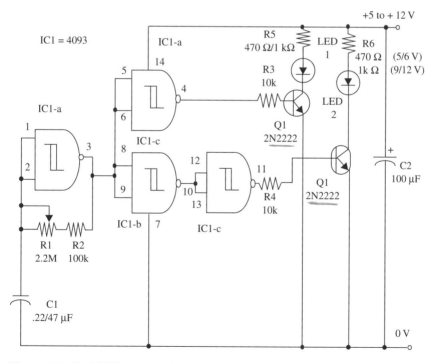

Figure 3.9 Dual LED flasher using transistor to drive several LEDs.

Parts List: Dual-LED Flasher

IC1	4093 CMOS integrated circuit
Q1, Q2	2N2222 NPN general purpose silicon transistors
LED1, LED2	Bicolor or red and green common LEDs, Radio Shack 276-012
R1	2,200,000 Ω potentiometer
R2	100,000 Ω, 1/4 W, 5% resistor
R3, R4	10,000 Ω, 1/4 W, 5% resistors
R5, R6	1/4 W, 5% resistor, values according power-supply voltage (see table included in the schematic diagram)
C1	0.22 or 0.47 µF ceramic or metal film capacitor
C2	100 µF, 16 WVDC electrolytic capacitor

Project 45: Complementary-Transistor, Dual-LED Flasher II (E)

This circuit drives two common LEDs or a bicolor LED with a flash rate that can be adjusted within a wide frequency range. The frequency range can be adjusted by R1 between 0.1 and 5 Hz.

Each transistor driver loads up to 100 mA; thus, you can wire several LEDs to the collectors, each one with the proper series resistor or small incandescent lamp, rated according the power supply voltage employed.

C1 can be altered to change the frequency range. Values between 0.22 and 2.2 µF can be used experimentally. R5 and R6 depends on the supply voltage, and the principal values are given in the schematic diagram.

A schematic diagram of the Complementary-Transistor Dual-LED Flasher is shown in Fig. 3.10.

Proper positioning of the polarized components (LEDs, electrolytic capacitors, and transistors) must be observed. Lamps such as the #40 (Radio Shack 272-1128) or two-pin 12 V, 40 mA (Radio Shack 272-1154) can replace the LEDs and series resistors in this project.

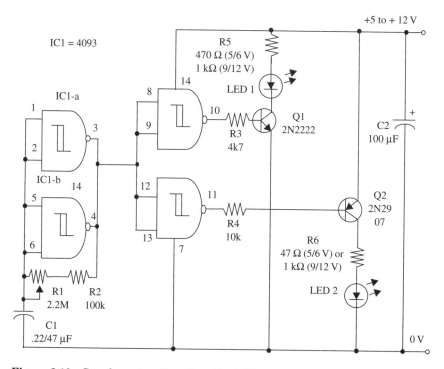

Figure 3.10   Complementary-Transistor Dual-LED Flasher.

Parts List: Complementary-Transistor Dual LED-Flasher

IC1	4093 CMOS integrated circuit
Q1, Q2	2N2222 NPN general purpose silicon transistors
LED1, LED2	Bicolor or green and red common LEDs, Radio Shack 276-012 or equivalent
R1	2,200,000 Ω potentiometer
R2	100,000 Ω, 1/4 W, 5% resistor
R3, R4	4,700 Ω, 1/4 W, 5% resistors
R5, R6	1/4 W, 5% resistors, values according power-supply voltage (see schematic diagram)
C1	0.22 or 0.47 µF ceramic or metal film capacitor
C2	100 µF, 16 WVDC electrolytic capacitor

Project 46: Two-Color LED Flasher III (E) (P)

This version of a two-color LED flasher uses two of the 4093's gates as low-frequency oscillators and other two gates as inverters to drive, in a complementary way, two LEDs or a bicolor LED.

The flash rate is adjusted by R1 and can range from 0.1 to 5 Hz. Values of C1 can be altered in the range between 0.22 and 2.2 µF to change the frequency range.

Rx depends on the power-supply voltage, and values are given in a table within the schematic diagram.

A schematic diagram of Two-Color LED Flasher III is shown in Fig. 3.11.

Parts List: Two-Color LED Flasher III

IC1	4093 CMOS integrated circuit
LED1, LED2	Two-color or two common LEDs, Radio Shack 276-012 or equivalent
R1	2,200,000 Ω potentiometer
R2	100,000 Ω, 1/4 W, 5% resistor
Rx	1/4 W, 5% resistor (see schematic diagram for values)
C1	0.22 or 0.47 µF ceramic or metal film capacitor
C2	100 µF, 16 WVDC electrolytic capacitor

Proper positioning of the polarized components (LEDs and C2) must be observed. The device can be used as component of alarms or in games, toys, and other applications.

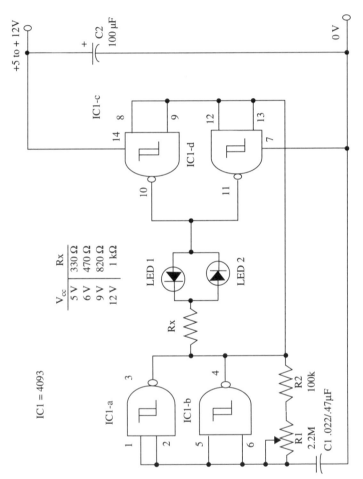

Figure 3.11   Two-Color LED Flasher III.

Project 47: Dual-Power Flasher (E) (P)

This dual flasher can drive incandescent lamps up to 1 A and can be used as part of warning systems in cars or trailers and in other applications. The circuit can be powered from supplies ranging from 6 to 12 V.

Darlington power transistors are used in the output stage. The complementary pair drives two lamps alternately in a flash rate that is controlled by potentiometer R1. The duty cycle is 50 percent, and the frequency range is between 0.1 and 5 Hz.

C1 can be altered, as in the other projects, to change the frequency range. Values up to 2.2 µF can be used experimentally, depending on the intended applications.

A schematic diagram of the flasher is shown in Fig. 3.12.

The transistors must be mounted on heatsinks. L1 and L2 should have voltages rates according power-supply voltage. With 12 V power supplies, you can use type #93 (Radio Shack 272-1116) or type PR-16 (250 mA, Radio Shack 272-1165).

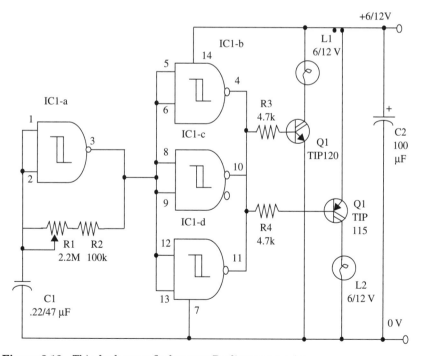

Figure 3.12   This dual-power flasher uses Darlington transistors.

Parts List: Dual-Power Flasher

IC1	4093 CMOS integrated circuit
Q1	TIP120 NPN Darlington power transistor
Q2	TIP115 PNP Darlington power transistor
L1, L2	200 mA to 1 A lamps (see text)
R1	2,200,000 Ω potentiometer
R2	100,000 Ω, 1/4 W, 5% resistor
R3, R4	4,700 Ω, 1/4 W, 5% resistors
C1	0.22 to 0.47 µF ceramic or metal film capacitor
C2	100 µF, 16 WVDC electrolytic capacitor

Project 48: Delayed Turn-Off Lamp (E) (P)

This circuit will turn off an incandescent lamp after a time delay in the range of a few seconds to a few minutes. Times are determined by C1 and can be extended up to half an hour with a 2,200 µF electrolytic capacitor.

When the power is on, the lamp is powered on, at which time C1 begins to charge through R1 until V_p is reached. Then, the IC1-a output goes high, and the driver stages formed by IC1-b, c, and d have their outputs drop to the low level. Thus, the output transistor is cut off, and L1 turns off.

The circuit can be powered from 6 to 12 V supplies. The lamp selection is according this voltage, and the rated current must be in the range of 100 mA to 1 A.

The circuit can be used in an automatic illumination system and in many other applications. To extend its power capabilities, you can replace the lamp with a 6 or 12 V relay.

A schematic diagram of the Delayed Turn-Off Lamp is shown in Fig. 3.13.

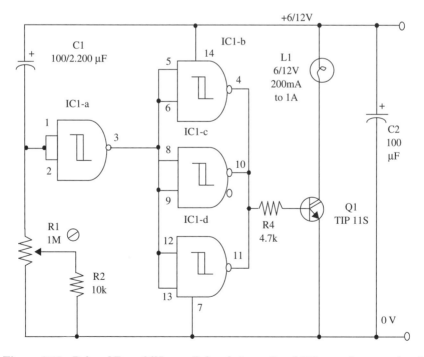

Figure 3.13 Delayed Turn-Off Lamp. Delays between 2 and 200 seconds are produced by this circuit.

Parts List: Delayed Turn-off Lamp

IC1	4093 CMOS integrated circuit
Q1	TIP120 NPN Darlington power transistor
L1	6 or 12 V incandescent lamp (see text)
R1	1,000,000 Ω potentiometer
R2	10,000 Ω, 1/4 W, 5% resistor
R3	4,700 Ω, 1/4 W, 5% resistor
C1	100 µF/220 µF, 16 WVDC electrolytic capacitor (see text)
C2	100 µF, 16 WVDC electrolytic capacitor

Transistor Q1 must be mounted on a heatsink. Several types of lamps are suitable for use in this project. See Project 47 for suggestions. To convert this circuit to a turn-on lamp, switch the position of C1 with the combination of R1/R2.

Project 49: Neon-Lamp Flasher (E)

This simple experimental circuit will flash a neon lamp at a rate of about 1 Hz. The frequency can be altered easily by changing C2 or, if you prefer, replacing R2 by a 2.2MΩ potentiometer with a 100 kΩ series resistor.

The circuit can be powered from supplies ranging from 5 to 12 V. The duty cycle is 50 percent and, to maintain the lamp "on" during large output pulses, a modulation is applied from IC1-a, which works as a low-frequency oscillator.

If you are supplying the unit from 9 V or greater power supplies, transistor Q1 must be mounted on a heatsink.

Transformer T1 is a small, non-critical power transformer with a 117 Vac primary and a 6 to 12 Vdc secondary. Secondary current can be rated from 100 mA to 300 mA. Perform experiments with the power transformers you have on hand.

The neon lamp is an NE-2 or NE-2H, and a series resistor is optional. Depending on the desired light level, you can perform experiments with 1 to 100 kΩ resistors. A schematic diagram of the Neon-Lamp Flasher is shown in Fig. 3.14.

Parts List: Neon Lamp Flasher

IC1	4093 CMOS integrated circuit
Q1	TIP31 NPN Power silicon transistor
T1	Power-transformer (see text), Radio Shack 273-1385, 12.6 V, 300 mA is suitable
NE-1, NE-2, or NE-2H	Neon lamp, Radio Shack 272-1101 or 272-1102
R1	47,000 Ω, 1/4 W, 5% resistor
R2	2,200,000 Ω, 1/4 W, 5% resistor
R3	4,700 Ω, 1/4 W, 5% resistor
C1	0.22 µF ceramic or metal film capacitor
C2	0.47 µF ceramic or metal film capacitor
C3	100 µF, 16 WVDC electrolytic capacitor

Components are placed on a solderless board or printed-circuit board except the transformer and neon lamp. Transistor Q1 must be mounted on a small heatsink.

Small, 4 to 15 W fluorescent lamps can be powered from this circuit. Perform some experiments with the ones that no longer function on the ac power line.

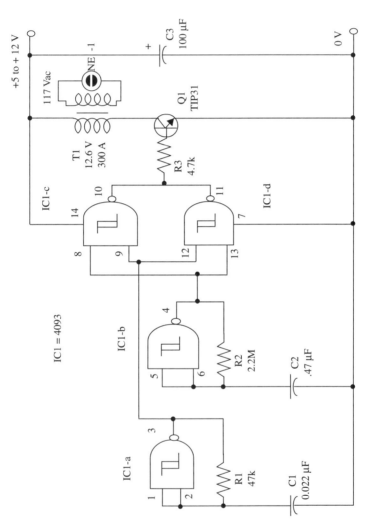

Figure 3.14 Neon Lamp Flasher. High voltage is produced by a transformer in this project.

Project 50: Delayed Turn-Off Flasher (E) (P)

This circuit can be used as part of a warning system to flash a lamp during a definite time delay after the power is turned on. The time delay can be adjusted to intervals between a number of seconds to minutes, but the circuit can be altered to extend this range. All you have to do is change C1 by units ranging from 47 to 1,500 µF.

The circuit can power lamps from 200 mA to 1 A and is supplied from 6 to 12 V supplies, depending on the lamp you intend to use.

The flash rate is given by R3 and C2, and the rate is about 1 Hz. You can also change this rate varying the value of C2 or replacing R3 with a 2.2 MΩ potentiometer with a 100 kΩ series resistor.

To add a start switch, place an SPST switch in series with C1. When this switch is turned on, capacitor C1 begins to charge, and IC1-b the oscillator is turned on. The lamp is then powered, producing flashes of light. When C1 is charged, IC1-a output goes low, IC1-b oscillator stops, and the lamp goes out.

A schematic diagram of the Delayed Turn-Off Flasher is shown in Fig. 3.15.

Parts List: Delayed Turn-Off Flasher

IC1	4093 CMOS integrated circuit
Q1	TIP120 NPN Darlington power transistor
L1	200 mA to 1 A incandescent lamp, 6 to 12 V (see text)
R1	1,000,000 Ω potentiometer
R2	10,000 Ω, 1/4 W, 5% resistor
R3	2,200,000 Ω, 1/4 W, 5% resistor
R4	4,700 Ω, 1/4 W, 5% resistor
C1	100 µF/220 µF, 16 WVDC electrolytic capacitor
C2	0.22 or 0.47 µF ceramic or metal film capacitor
C3	100 µF, 16 WVDC electrolytic capacitor

Suggestions for the lamps are given in Project 47. Transistor Q1 must be mounted on a heatsink. The proper position of the polarized components must be observed.

You can increase the power output replacing the output transistor by using any power FET with current ratings of 2 A or more.

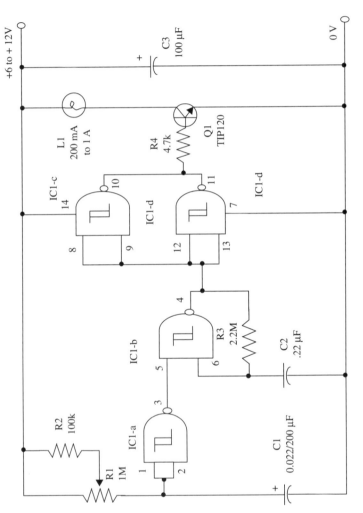

Figure 3.15 Delayed Turn-Off Flasher. P1 adjusts time delay.

Project 51: Delayed Turn-On Flasher (E) (P)

With this device, the lamp begins to flash from a number of seconds to minutes after the power is on. The circuit can be used as part of alarms, warning systems, home appliances, and so on.

As with the previous projects, the time delay and flash rate can be altered by changing the values of capacitors C1 and C2. Transistor Q1 must be mounted on a heatsink, and the lamps used are the same suggested in Project 47.

A schematic diagram of the Delayed Turn-On Flasher is shown in Fig. 3.16.

Parts List: Delayed Turn-On Flasher

IC1	4093 CMOS integrated circuit
Q1	TIP120 NPN Darlington transistor
L1	200 mA to 1 A lamp, 6 or 12 V (see text)
R1	1,000,000 Ω potentiometer
R2	10,000 Ω, 1/4 W, 5% resistor
R3	2,200,000 Ω, 1/4 W, 5% resistor
R4	4,70 Ω, 1/4 W, 5% resistor
C1	100 μF/220 μF, 16 WVDC electrolytic capacitor
C2	0.22/0.47 μF ceramic or metal film capacitor
C3	100 μF, 16 WVDC electrolytic capacitor

For a delayed turn-on, intermittent relay, replace L1 with a 6 V or 12 V, 100 mA to 500 mA coil relay. Contacts must be selected according the load you intend to control.

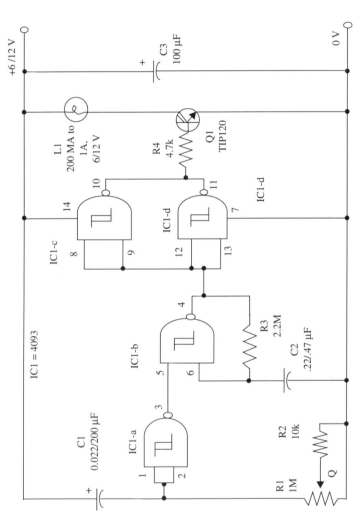

Figure 3.16 Delayed Turn-On Flasher. The lamp will begin to flash after a time delay from a few seconds to minutes.

Project 52: Touch-Triggered LED Flasher (E)

This experimental circuit will flash an LED when you touch the sensor with your fingers. Applications include alarms and experiments in science for high school students and amateur scientists.

The circuit can also be used in games and toys, and the power supply can range from 5 to 12 V. Rx depends on the power-supply voltage and is given in a table in the schematic diagram.

Sensitivity is determined by R1, which can be reduced to values down to 1 MΩ, and the flash rate is determined by R2 in a range from 0.1 to 5 Hz. Transistorized output stages can be added to drive lamps or other powerful loads.

A schematic diagram of the Touch-Triggered LED-Flasher is shown in Fig. 3.17.

Parts List: Touch-Triggered LED-Flasher

IC1	4093 CMOS integrated circuit
LED	Common red, yellow, or green LED
X1	Sensor (see text)
R1	10,000,000 Ω, 1/4 W, 5% resistor
R2	2,200,000 Ω, 1/4 W, 5% resistor
R3	100,000 Ω, 1/4 W, 5% resistor
Rx	1/4 W, 5% resistor, value according the power-supply voltage (see text)
C1	0.22 µF or 0.47 µF ceramic or metal film capacitor
C2	100 µF, 16 WVDC electrolytic capacitor

The sensor is formed by two plates or screws placed together, to be touched at same time with your fingers. Positions of the polarized components, such as the LEDs and C2, must be observed.

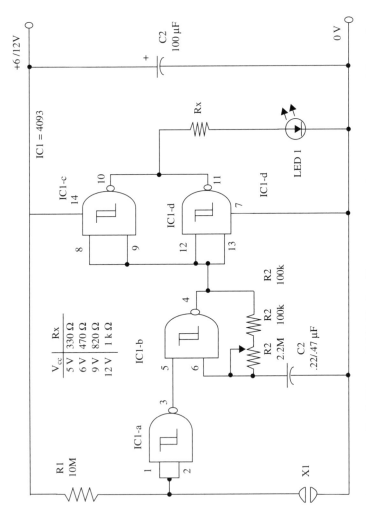

Figure 3.17   Touch-Triggered LED Flasher. You can use a transistor output stage to drive a relay.

Project 53: Variable Duty Cycle Lamp Flasher (E) (P)

The interval between flashes can be adjusted within a large range of values with this circuit. The flash duration is fixed, but you can control it replacing R3 with a 2.2 MΩ potentiometer with a 100 kΩ series resistor.

The device can be used as part of alarms, warning systems, games, toys, decorations, cars, and so forth. The power supply depends on the lamp and can be a 6 or 12 V unit. Output current is up to 500 mA, depending on the lamp. The operational principle is the same as with the other variable duty cycle projects described in this book.

A schematic diagram of the Variable Duty Cycle Lamp Flasher is shown in Fig. 3.18.

Parts List: Variable Duty Cycle Lamp-Flasher

IC1	4093 CMOS integrated circuit
Q1	TIP31 NPN power transistor
D1, D2	1N4148 general purpose silicon diodes
L1	200 to 500 mA, 6 or 12 V lamp (see text)
R1	2,200,000 Ω potentiometer
R2	100,000 Ω, 1/4 W, 5% resistor
R3	2,200,000 Ω, 1/4 W, 5% resistor
R4	2,200 Ω, 1/4 W, 5% resistor
C1	0.22µF/0.47 µF ceramic or metal film capacitor
C2	100 µF, 16 WVDC electrolytic capacitor

The lamps are described in Project 47, and transistor Q1 must be mounted on a heatsink. The proper position of the polarized components must be observed.

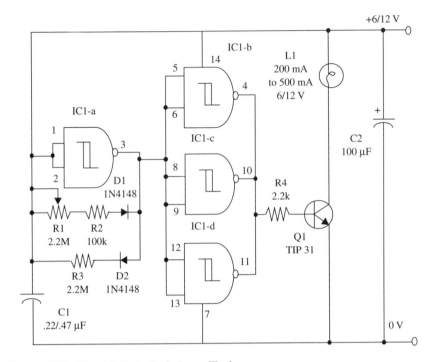

Figure 3.18   Variable Duty Cycle Lamp Flasher.

Project 54: Dark-Activated Lamp Flasher (P)

This circuit can be used in warning systems, driving an incandescent 6 V or 12 V lamp. The circuit works as follows.

When light is striking sensor Q1, IC1-a output is low, and the IC1-b oscillator is off. Thus, IC1-b is high, which results in a low IC1-c output and IC1-d output. Transistor Q2 is cut, and the lamp is off. When light drops off, IC1-a output goes high, and IC1-b oscillator turns on, driving the output transistor via IC1-c and IC1-d.

Transistor Q2 can drive lamps up to 500 mA and must be mounted on a heatsink. Sensitivity is adjusted by R1, and flash rate is adjusted by R2.

You can replace the potentiometer with a CdS photoresistor. In this case, also replace R1 with a 100 kΩ potentiometer.

A schematic diagram of the Dark-Activated Lamp Flasher is shown in Fig. 3.19.

Parts List: Dark-Activated Lamp Flasher

IC1	4093 CMOS integrated circuit
Q1	TIL414 or equivalent phototransistor, Radio Shack 276-145
Q2	TIP120 NPN Darlington power transistor
L1	200 mA to 1 A, 6 or 12 V lamp (see text)
R1	2,200,000 Ω or 4,700,000 Ω trimmer potentiometer
R2	100,000 Ω, 1/4 W, 5% resistor
R3	2,200,000 Ω potentiometer
R4	100,000 Ω, 1/4 W, 5% resistor
R5	4,700 Ω, 1/4 W, 5% resistor
C1	0.22 or 0.47 µF ceramic or metal film capacitor
C2	100 µF, 16 WVDC electrolytic capacitor

Transistor Q1 must be mounted on a heatsink. Observe proper positioning of polarized components such as transistors and the electrolytic capacitor. Q1 is mounted inside a tube and, for a more directional performance, you can include a convergent lens in the front of the tube.

To operate the unit, adjust the sensitivity using R1 to the dark level required to trigger the circuit.

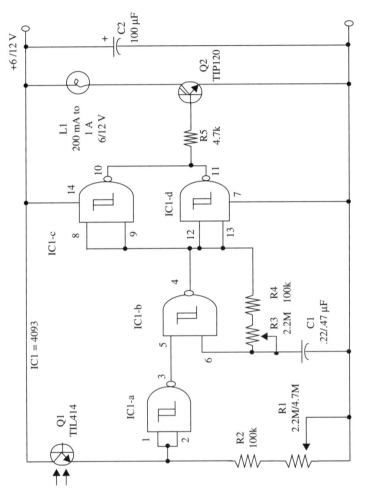

Figure 3.19 Dark-Activated Lamp Flasher.

Project 55: Variable Duty Cycle Flasher (E) (P)

This circuit provides a variable-length pulse, driving an incandescent lamp rated from 100 mA to 1 A. The unit can be used in warning systems, alarms, and many other applications.

The frequency can be adjusted within a wide range of values using a potentiometer. The circuit operates as follows.

IC1-a is a free-running, low-frequency oscillator operating in a frequency range between 0.1 and 1 Hz. Oscillator output pulses are applied simultaneously to the IC1-d input and to an RC network formed by R3, R4, and C2. The time constant of the network will delay the output pulse to be applied to pin 14 of IC1-d. Thus, according this delay, the output pulse of IC1-d and the duration of the lamp flash will be altered.

If we have small delay, the two IC1-d input pulses arrive at practically the same time, and the duty cycle of the lamp flash will be the same as the free-running oscillator formed by IC1—about 50 percent. But, by adjusting R3, we can trigger IC1-d with large delays. If the delay is long enough to trigger IC1-d at the end of the pulse, we will get low-duration flashes or small duty cycles.

D1 is used to discharge capacitor C2 at the end of each pulse. When IC1-a output goes low, between flashes, D1 is directly biased, causing C1 to discharge through it.

A Darlington power transistor is used in the output stage to drive lamps up to 1 A. The power supply can be a 6 or 12 V unit, depending on the lamp.

The device can be used as part of warning systems, alarms, decorations, and so forth. C1 and C2 can be altered to change the frequency and duty cycle, but the values should be close as possible.

A schematic diagram of the device is shown in Fig. 3.20.

Transistor Q1 must be mounted on a heatsink. The lamp can be the same one suggested in Project 47. A suitable base for the lamp should be used.

The transistor can be replaced by a power FET to get better performance. Any "IRF" power FET with drain currents rated at 2 A or more can be used in this project.

You can replace the lamp with a 6 or 12 V relay to control external loads, including common 117 Vac lamps.

Parts List: Variable Duty Cycle Flasher

IC1	4093 CMOS integrated circuit
Q1	TIP120 Darlington power transistor
D1	1N4148 general purpose silicon diode
L1	100 mA to 1 A, 6 V or 12 V lamp (see text)
R1, R3	2,200,000 Ω potentiometer
R2, R4	100,000 Ω, 1/4 W, 5% resistors
R5	4,700 Ω, 1/4 W, 5% resistor
C1, C2	0.47 or 1 µF ceramic or metal film capacitor
C3	100 µF, 16 WVDC electrolytic capacitor

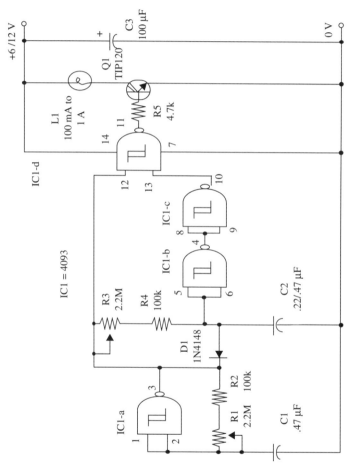

Figure 3.20 Variable Duty Cycle Flasher. Frequency is adjusted by R1, and duty cycle by R2.

4

Time-Delay Projects

The 4093 can be readily coupled to electronic time-delay circuits and used in a wide range of applications for home, industry, robotics, cars, scientific experiments, school science works, ad infinitum. The circuits in this chapter can be made to function in either the delayed turn-on or delayed turn-off modes, and they can drive either ac or dc loads, LED circuits, buzzer circuits, home appliances, and so forth in a wide power range.

Many circuits are experimental and also can be used as part of more complex projects, but many of these circuits are complete and can be used as shown. In the last part of this book, we will provide other projects combining time-delay circuits, flashers, buzzers, output stages, and many configurations shown in a separate form throughout this book.

Project 56: Simplest Timer (E) (P)

This is the simplest configuration of a timer using the 4093 IC. After the adjusted time delay, the LED can turn on or off according the reader's preferred version. The time delay can be adjusted in the range between some seconds and more than half an hour, and the circuit can be used as an egg timer or parking timer, in games (e.g., chess), and in many other applications.

The project can be powered from four AA cells or a 9 V battery. Current drain depends on the state of the LED.

With the LED off, current drain is only about 1 mA. With the LED on, current drain is between 10 and 30 mA according to the power-supply voltage and R3.

If you want a turn-off action for the LED, wire it as shown by the continuous line in the schematic diagram. If you want a turn-on action, wire the LED as shown by the interrupted line in the schematic diagram.

The circuit works as follows. When S1 is closed, the IC1-a output is low, and C1 begins to charge through R1 and R2 until V_n is reached (pin 2). At the same time, IC1-b, c, and d outputs are high, and the LED is on.

When V_n is reached, the IC1-a output goes high, and this logic level is applied to IC1-b, c, and d inputs. The three gates' outputs, initially at the high level, pass to the low level, turning off the LED. If the LED is wired to the positive power line, this component, initially off when S1 is closed, turns on after the adjusted time delay.

A schematic diagram of the Simplest Timer is shown in Fig. 4.1.

The time delay depends on the capacitor C1. With a 470 µF capacitor and a 4.7 MΩ potentiometer, the maximum time delay is about half an hour. Capacitors up to 1,000 µF can be used for longer time delays.

You can also mount your timer on a printed circuit board and house it into a plastic box to provide a portable version.

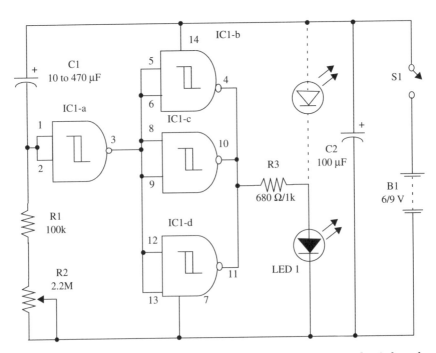

Figure 4.1 Simplest Timer. The LED connection depends on the desired switch mode.

Parts List: Simplest Timer

IC1	4093 CMOS integrated circuit
LED1	Red common LED
S1	SPST switch
R1	6 or 9 V, four AA cells or 9 V battery
R1	100,000 Ω, 1/4 W, 5% resistor
R2	2,200,000 Ω or 4,700,000 Ω potentiometer
R3	680 Ω, 1/4 W, 5% resistor (6 V supply), 1,000 Ω, 1/4 W, 5% resistor (9 V supply)
C1	10 µF to 470 µF, 16 WVDC electrolytic capacitor
C2	100 µF, 12 WVDC electrolytic capacitor

Project 57: Auto Turn-Off Relay (E) (P)

This circuit is designed to apply power to a load as soon as an operating switch is closed and to remove the power automatically after a preset time delay.

Time delays up to half an hour can be achieved with this project, which can be used to turn off a TV, audio equipment, lamps, and many other home appliances. The basic project is powered directly from a 12 V or a 6 V battery, but modifications can be made to power it from the ac power line as shown in Fig. 4.2.

We recommend a DPDT mini relay (Radio Shack 275-249) that can be mounted on a solderless board or a universal printed circuit board, but other sensitive relays with coils rated for 6 or 12 V can be used. The relay's coils should have resistances in the range between 200 to 500 Ω (12 V) or 100 to 250 Ω (6 V).

Circuit operation is the same as described in Project 56. The only difference is the output stage with a transistor to drive the relay.

A schematic diagram of the Auto Turn-Off Relay is shown in Fig. 4.3.

Parts List: Auto Turn-Off Relay

IC1	4093 CMOS integrated circuit
Q1	2N2222 NPN general purpose silicon transistor
D1	1N4148 or equivalent general purpose silicon diode
K1	12 Vdc, 43 mA, 280 Ω mini DPDT relay, contacts rated to 1 A (Radio Shack 275-249) or a 6 V unit (see text)
LED1	Red common LED
S1, S2	Mini SPST momentary switch normally open, Radio Shack 275-1547
R1	100,000 Ω, 1/4 W, 5% resistor
R2	2,200,000 Ω or 4,700,000 Ω potentiometer
R3	4,700 Ω, 1/4 W, 5% resistor
R4	680 Ω, 1/4 W, 5% resistor (6 V) 1,000 Ω, 1/4 W, 5% resistor (12 V)
C1	10 µF to 1,000 µF, 12 WVDC electrolytic capacitor (see text)

Proper positioning of the polarized components (diode, electrolytic capacitors, and power supply) must be observed. Wiring to the load must be rated according the amount of current drain.

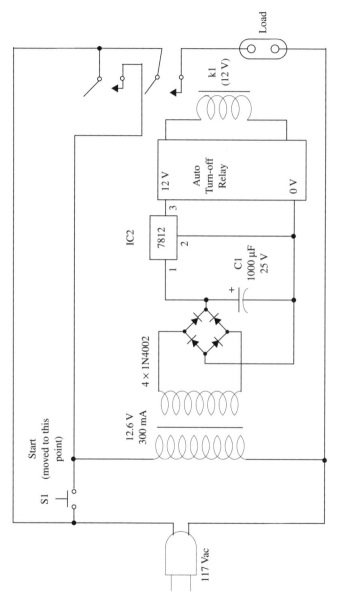

Figure 4.2 Modification to control a load directly from the ac power line. Note the new position of S1.

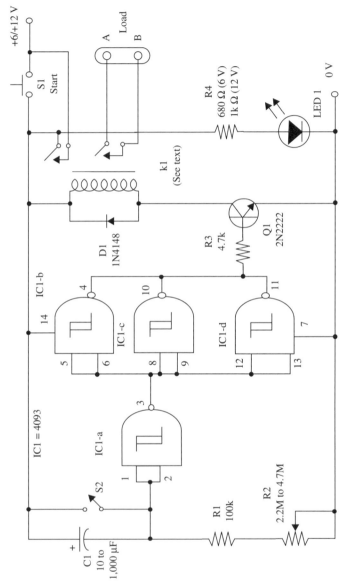

Figure 4.3   Auto Turn-Off Relay.

Capacitor C1 is chosen according the desired delay range. Using a 1,000 µF capacitor and a 4.7 MΩ potentiometer, time delays up to half an hour can be obtained.

To operate the unit, connect the load to the relay contacts and adjust R2 to the desired time delay. Press S1 to start. At this time, the load receives power, and LED1 glows.

After the adjusted time delay, the unit automatically turns off, along with the controlled load. To use the unit again, you have to press S2 to discharge C1 before a new start.

Note: As the 4093 has a very high input impedance, all projects with large values of timer capacitors should have a parallel SPST momentary switch to discharge them after each cycle of operation, except in the versions that use automatic discharge networks.

Project 58: Simple Timer II (E) (P)

Small home appliances and circuits can be turned on after a time de-lay ranging from seconds to minutes with this simple timer. The unit functions as the two previous timers we have described in this section, with the difference that it uses a PNP transistor in the output stage.

Using a PNP transistor, the relay is powered when IC1-b, c, and d outputs go to the low level at the end of the adjusted time delay.

You can use this timer as part of other, more complex projects or as an easy-to-use timer to control appliances from the ac power line, add-ing some components as shown in Fig. 4.4. Using a 1 A relay (e.g., Ra-dio Shack mini SPST relay), you can control ac loads up to 100 W.

The circuit can be powered by AA cells or, if you prefer, from an ac power supply. (See in the introduction of this book for some sugges-tions with regard to power supplies.)

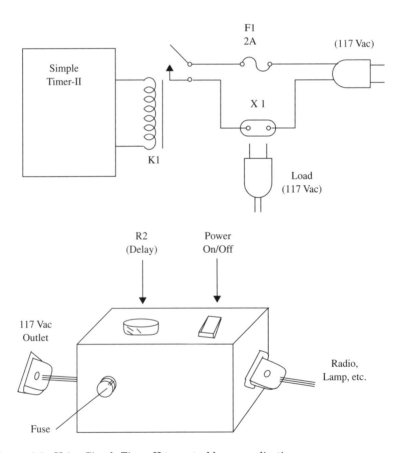

Figure 4.4   Using Simple Timer II to control home applications.

A schematic diagram of the Simple Timer II is shown in Fig. 4.5.

Parts List: Simple Timer II

IC1	4093 CMOS integrated circuit
Q1	2N2907- NPN general purpose silicon transistor
D1	1N4148 general purpose silicon diode
K1	12 Vdc, 43 mA, 280 Ω mini DPDT or SPST relay, contacts rated to 1 A (Radio Shack 275-249 or equivalent) or a 6 V unit (see text)
R1	100,000 Ω, 1/4 W, 5% resistor
R2	2,200,000 Ω potentiometer
R3	4,700 Ω, 1/4 W, 5% resistor
C1	10 µF to 1,000 µF, 12 WVDC electrolytic capacitor (see text)
C2	100 µF, 12 or 16 WVDC electrolytic capacitor

The proper position of the polarized components (diode and electrolytic capacitors) must be observed.

To operate the unit, set the time delay by R2 and turn on the power supply. After the adjusted time delay, the relay will be energized, acting on the load.

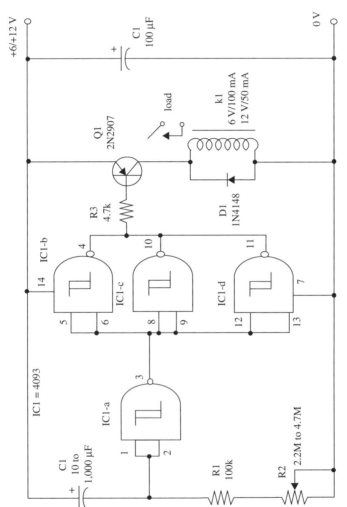

Figure 4.5 Simple Timer II. See modifications to operate the unit from the ac power line.

Project 59: Continuous-Sound Output Timer (E) (P)

This circuit produces a continuous sound after the adjusted time delay. The device can be used as a darkroom timer, egg timer, to control printed circuit etching, in games, and many other applications. Time delays can be adjusted in a range from minutes to half an hour.

A compact unit can be built and housed in a small plastic box. You can fit this unit in your pocket for portability and convenience.

The circuit works like the previous projects in this part of the book, so the only explanation necessary is about the output stages. IC1-b acts as a control LED audio oscillator, running only at the end of the adjusted time delay. This oscillator supplies the output signal to a drive stage, and afterward to a piezoelectric tran.3ducer.

Time delay is adjusted by R2, and the output tone is adjusted by R3. C1 determines the time range. With a 1,000 µF capacitor and a 4.7 MΩ potentiometer, we can get time delays of up to half an hour. The circuit can be powered from AA cells or 9 V battery.

A schematic diagram of the timer is shown in Fig. 4.6.

Parts List: Continuous Sound Output Timer

IC1	4093 CMOS integrated circuit
X1	Piezoelectric transducer or crystal earpiece, Radio Shack 273-072 or equivalent
S1	SPST toggle or slide switch
S2	SPST momentary switch
B1	6 V or 9 V (four AA cells or battery) and holder
R1	100,000 Ω, 1/4 W, 5% resistor
R2	2,200,000 Ω or 4,700,000 Ω potentiometer
R3	100,000 Ω potentiometer or trimmer
R4	10,000 Ω, 1/4 W, 5% resistor
C1	10 µF to 1,000 µF, 16 WVDC, electrolytic capacitor (see text)
C2	0.022 µF ceramic or metal film capacitor
C3	100 µF, 12 WVDC electrolytic capacitor

R2 and R3 can be replaced by trimmer potentiometers or common resistors if you want a definite time delay and output tone. S2 is required to discharge C1 after each period of use as the capacitor maintains its charge during long time delays. X1 is a piezoelectric transducer or a crystal earpiece.

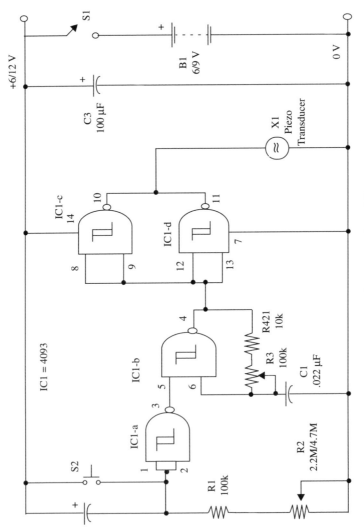

Figure 4.6 Continuous-Sound Timer.

To use the timer, adjust the time delay via R2 and turn on the power switch, closing S1. After the desired time delay, the buzzer will produce a continuous sound with a tone given by R3 adjustment. R3 and R4 can be replaced by a 27 or 33 kΩ common resistor.

Project 60: Pulsed-Tone Timer (E) (P)

You can use this device as an egg timer or darkroom timer, in games, and in many other applications, as in the case of the preceding projects in this chapter. This circuit will produce an intermittent tone after the adjusted time delay. Time delays can be adjusted from seconds to more than half an hour.

The circuit can be powered from four AA cells (6 V) or a battery (9 V), and current drain is very low—a few milliamperes at any condition (tone or off). The circuit can also be mounted on an universal printed circuit board and housed in a small plastic box for portable use. This version can be used anywhere and transported in your pocket.

A schematic diagram of the pulsed tone timer is given in Fig. 4.7.

Parts List: Pulsed-Tone Timer

IC1	4093 CMOS integrated circuit
X1	Piezoelectric transducer or crystal earphone, Radio Shack 272-073 or equivalent
S1	SPST toggle or slide switch
S2	SPST momentary switch
B1	6 or 9 V (four AA cells or battery)
R1	100,000 Ω, 1/4 W, 5% resistor
R2	2,200,000 Ω or 4,700,000 Ω potentiometer
R3	39,000 Ω, 1/4 W, 5% resistor
R4	1,500,000 Ω, 1/4 W, 5% resistor
C1	10 µF to 1,000 µF, 12 WVDC electrolytic capacitor (see text)
C2	0.022 µF ceramic or metal film capacitor
C3	0.22 µF or 0.47 µF ceramic or metal film capacitor
C4	100 µF, 12 WVDC electrolytic capacitor

Proper positioning of the polarized components (electrolytic capacitors and battery) must be observed. Operation of the timer is very easy: first adjust time delay via R2. Then, close S1 and wait. After the adjusted time, the device will produce an intermittent tone. To stop the tone, open S1.

To use the unit again, first press S2 to discharge C1, then close S1 to set a new time delay. With a 1,000 µF capacitor (C1) and a 4,700,000 Ω potentiometer, the time delay ranges from seconds to more than half an hour.

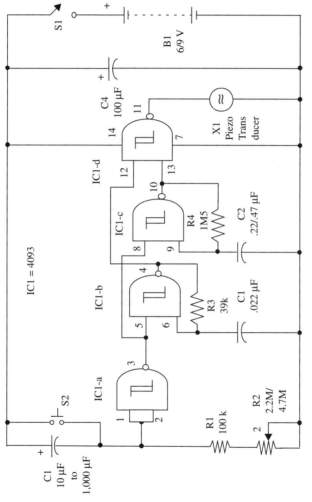

Figure 4.7 Pulsed-Tone Timer.

Project 61: Turn-Off Timer (E) (P)

This simple timer will turn off any load connected to its output after a time delay that can be adjusted from seconds to more than half an hour. You can use this device to turn off heaters, electric tools, darkroom enlarger lamps, and many other appliances. Capacitor C1 is chosen according the intended application. Short time delays can be obtained with low values, in the range between 10 and 100 µF. Large values (e.g., a 1,000 µF capacitor and a 4.7 MΩ potentiometer) can give time delays as high as 45 minutes.

Current drain in the load is fixed by the relay contacts' rates. With a 1 A relay, you can control loads up to 100 W. If you want to control more powerful loads, a heavy-duty relay must be used. The relay coil can be rated for 6 or 12 V. The recommended transistor can sink currents up to 100 mA, and this determines the coil characteristics.

A schematic diagram of the Turn-Off Timer is shown in Fig. 4.8.

Parts List: Turn-Off Timer

IC1	4093B CMOS integrated circuit
Q1	2N2222 NPN general purpose silicon transistor
D1	1N4148 general purpose silicon diode
K1	6 or 12 V relay (12 V, 43 mA, 280 Ω; Radio Shack 275-249 is a suitable unit)
S1	SPST momentary switch
R1	2,200,000 Ω or 4,700,000 Ω potentiometer
R2	100,000 Ω, 1/4 W, 5% resistor
R3	4,700 Ω, 1/4 W, 5% resistor
C1	10 µF to 1,000 µF, 12 WVDC electrolytic capacitor
C2	100 µF, 16 WVDC electrolytic capacitor

Proper positioning of the polarized components (diode, electrolytic capacitors, and transistor) must be observed.

Connections to external loads are made as shown in Project 57. To drive heavy-duty relays. you should replace Q1 with a TIP31 and reduce R3 to 2.2 kΩ. Relays up to 500 mA coils can be driven with this new configuration.

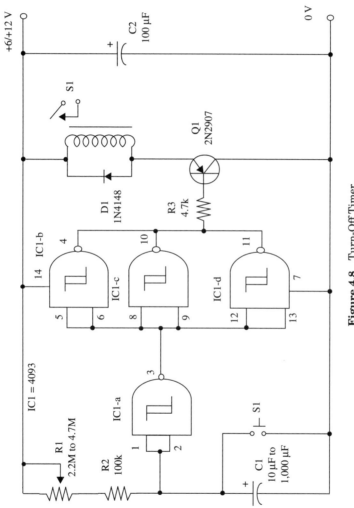

Figure 4.8 Turn-Off Timer.

Project 62: Pulsed-Tone Turn-Off Timer (E) (P)

This circuit produces a pulsed tone during an adjustable time delay. The circuit has some interesting uses such as in games, home applications, and others.

You can power the device with four AA cells or a 9 V battery. Current drain is very low, extending battery life to several months.

Time delays are in the range of a few seconds to more than half an hour. A piezoelectric transducer or crystal earphone produces a high-level intermittent tone. You can house all the components in a small plastic box to obtain a portable, easy-to-use unit.

The device has only one adjustment, R1, for the time delay. A schematic diagram for the Pulsed-Tone Turn-Off Timer is shown in Fig. 4.9.

Parts List: Pulsed-Tone Turn-Off Timer

IC1	4093 CMOS integrated circuit
X1	Piezoelectric transducer or crystal earpiece, Radio Shack 273-073 or equivalent
S1	SPST toggle or slide switch
S2	SPST momentary switch
R1	2,200,000 Ω or 4,700,000 Ω potentiometer
R2	100,000 Ω, 1/4 W, 5% resistor
R3	39,000 Ω, 1/4 W, 5% resistor
R4	2,200,000 Ω, 1/4 W, 5% resistor
C1	10 µF to 1,000 µF, 12 WVDC electrolytic capacitor (see text)
C2	0.022 µF ceramic or metal film capacitor
C3	0.47 µF ceramic or metal film capacitor
C4	100 µF, 12 WVDC electrolytic capacitor

Proper positioning of the electrolytic capacitors must be observed. R3 determines the tone, and R4 the interruption rate. Values of these components can be varied to change the sound.

S2 should be pressed to use the unit again. This switch discharges C1 after a time delay. With a 1,000 µF capacitor and a 4.7 µΩ potentiometer, time delays are up to 45 minutes.

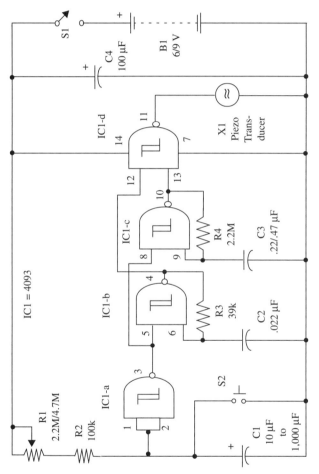

Figure 4.9 Pulsed-Tone Turn-Off Timer. X1 also can be a crystal earphone.

Project 63: Dual Turn-On Timer (E) (P)

This circuit will drive two relays after two different time delays. First, K2 is activated after a time delay adjusted by R2. Then, after a new time delay, adjusted by R5, K1 closes its contacts.

The circuit can be used as a sequential timer in several types of automatic systems. For instance, you can use this circuit as part of an "intelligent" alarm, first turning on a siren and then, after some minutes, a lamp. The second action of the circuit, turning on the lamp, suggests to the intruder the presence of another human in the vicinity.

The circuit can be powered from 6 or 12 V supplies. In alarm systems, you should use heavy-duty batteries or ac power supplies, because, when closed, the two relays drain an amount of current exceeds the capacity of small AA cells and other small batteries. With the dual time delays, you can obtain a total time delay of up to an hour and half.

A schematic diagram of the Dual Turn-On Timer is shown in Fig. 4.10.

Parts List: Dual Turn-On Timer

IC1	4093 CMOS integrated circuit
Q1	2N2222 NPN general purpose silicon transistor
Q2	2N2907 PNP general purpose silicon transistor
D1, D2	1N4148 general purpose silicon diodes
K1, K2	6 or 12 V relay, 100 mA coil (Radio Shack 275-249, 12 Vdc, 43 mA, 280 Ω, is a suitable unit)
S1, S2	SPST momentary switch
R1, R4	100,000 Ω, 1/4 W, 5% resistors
R2, R3	2,200,000 Ω or 4,700,000 Ω potentiometers
R5, R6	4,700 Ω, 1/4 W, 5% resistors
C1, C2	10 µF to 1,000 µF, 12 WVDC electrolytic capacitors (see text)
C3	100 µF, 16 WVDC electrolytic capacitor

Proper positioning of the polarized components (the two diodes and the electrolytic capacitors) must be observed. Using mini DPDT relays, it's easy to build a compact unit, as these components can be mounted directly on the solderless board or universal printed circuit board.

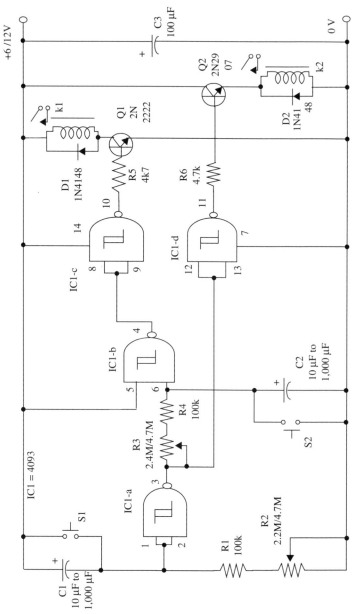

Figure 4.10 Dual Turn-On Timer.

To drive powerful relays, you can use other transistors in the output stage. Q1 can be replaced by a TIP31 to drive up to 500 mA relays, and Q2 can be replaced by a TIP32. These transistors must be mounted on heatsinks if the power supply voltage is 12 V.

Project 64: Turn-On and Turn-Off Timer (E)

As the name suggests, this timer will turn on a load after an adjustable time delay and turn off the same load after a second adjustable time delay. In this project, load is a simple LED, but you can easily alter it to drive other loads such as relays, lamps, motors, and so on.

The project can be used to demonstrate the monostable action of the two 4093 gates used in the basic configuration. The first time delay (turn on) can be adjusted from a few seconds to more than half an hour using R1. On time can be adjusted in the same range using R3.

A schematic diagram of the experimental Turn-On and Turn-Off Timer is shown in Fig. 4.11.

Parts List: Turn-On and Turn-Off Timer

IC1	4093 CMOS integrated circuit
LED1	Red common LED
R1, R4	1,000,000 Ω to 4,700,000 Ω potentiometer or trimmer potentiometers
R2	100,000 Ω, 1/4 W, 5% resistor
R3	47,000 Ω, 1/4 W, 5% resistor
R5	1/4 W, 5% resistor, according to power-supply voltage (see table in the schematic diagram)
C1, C2	10 µF to 1,000 µF, 12 WVDC electrolytic capacitor (see text)

The positions of the polarized component (electrolytic capacitor) must be observed. To produce an inverted action, you can wire the LED between pins 10 and 11 and the negative power line. The LED will turn off and, after a time delay, turn on again. With a 1,000 µF and a 4.7 µΩ potentiometer, the maximum time delay is up to 30 minutes.

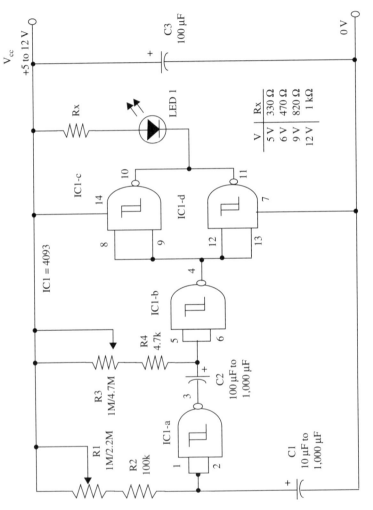

Figure 4.11   Turn-On and Turn-Off Timer.

Project 65: Turn-Off and Turn-On Timer (E)

This experimental dual timer turns off a load after a time delay and turns the same load on again after another time delay. Time delays in each case can reach more than 45 minutes, giving a total time delay of more than an hour and a half. In our experimental project, the load is an LED, but it easily can be altered to drive other loads.

The circuit consists of two monostable multivibrators made with two of the four 4093 gates. Capacitors C1 and C2, with the associated resistors (R1, R2, R3, and R4), are responsible for the time delays. With 1,000 µF capacitors and a 4.7 MΩ potentiometer, the maximum time delay is 45 minutes.

A schematic diagram of the Turn-On and Turn-Off Timer is given in Fig. 4.12.

Parts List: Turn-Off and Turn-On Timer

IC1	4093 CMOS integrated circuit
LED1	Red common LED
R1, R3	100,000 Ω, 1/4 W, 5% resistors
R2, R4	1,000,000 Ω to 4,700,000 Ω potentiometers
R5	1/4 W, 5% resistor (see table in the schematic diagram)
C1, C2	10 µF to 1,000 µF, 12 WVDC electrolytic capacitors
C3	100 µF, 16 WVDC electrolytic capacitor

Proper positioning of the polarized components (LED and electrolytic capacitors) must be observed.

In use, when the power is on (power switch closed), the LED glows during a time delay adjusted by R2. After this time delay, the LED turns off and remains at this state during a time delay determined by R4 adjustment. After the second time delay, the LED will turn on again. To set a new time delay, open the power switch and wait a few minutes to discharge C1 and C2.

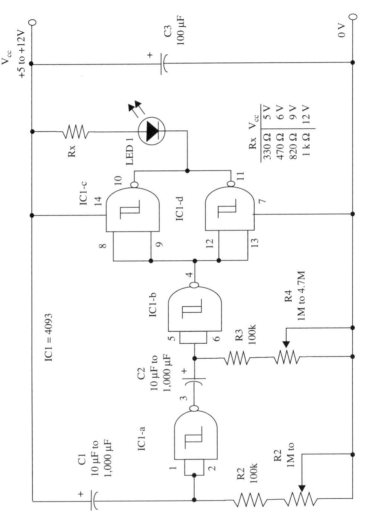

Figure 4.12 Turn-Off and Turn-On Timer.

Project 66: Turn-On and Turn-Off Timer with Relay (E) (P)

A relay is added to the basic Project 65 to offer a wider range of uses for the turn-on and off-timer. You can use this project as part of car alarms, with a second time delay to disconnect the horn, thereby avoiding battery discharge. You can adjust the turn-on time to a number of seconds and the turn-off time delay to a number of minutes, determining the time during which the horn will sound.

Circuit operation is as the other monostable projects described in this book, and the relay must be chosen according load requirements. A mini DPDT relay such as the Radio Shack 275-259 controls small appliances rated to up than 1 A. Heavy-duty relays should be used to control more powerful loads, but if their coils are rated to more than 100 mA to 500 mA, transistor Q1 should be replaced with a TIP32, and R5 by a 2.2 kΩ resistor.

A schematic diagram of this timer is shown in Fig. 4.13.

Parts List: Turn-On and Turn-Off Timer with Relay

IC1	4093 CMOS integrated circuit
Q1	2N2907 PNP general purpose silicon transistor
D1	1N4148 general purpose silicon diode
R1, R4	2,200,000 Ω to 4,700,000 Ω potentiometers
R2, R3	100,000 Ω, 1/4 W, 5% resistors
K1	6 or 12 V relay (see text), Radio Shack mini DPDT relay 275-249 is a suitable unit
C1, C2	10 µF to 1,000 µF, 12 WVDC electrolytic capacitors
C3	100 µF, 16 WVDC electrolytic capacitor

Proper positioning of the polarized components (diode, transistor, and electrolytic capacitors) must be carefully observed.

The mini DPDT relay is mounted on the solderless board as shown in the figure. If you're using an equivalent, be careful with the terminals' positions—they could be different.

The load is connected to the relay as shown in other projects in this timer series. R1 adjusts the first time delay, and R2 adjusts the second.

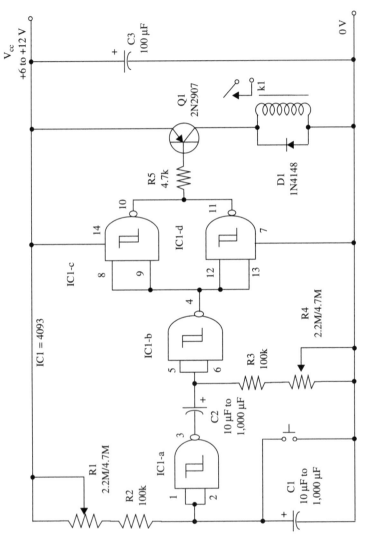

Figure 4.13   Turn-On and Turn-Off Timer with Relay.

Project 67: Dual Mini Timer (E) (P)

You can carry this timer with you in your pocket and use it as an egg timer, parking timer, darkroom timer, and in many other applications. The first time delay is adjusted via R1. After this time period, the piezoelectric transducer or crystal earphone will sound during a second time period, adjusted by R3. The first and second time delays can be adjusted in a range between a few seconds to up than 45 minutes, according the values of capacitors C1 and C2.

The circuit can be powered by four AA cells or a 9 V battery, and current requirements are very low, extending battery life to several weeks. The tone is determined by C3 and R5. To get an adjustable tone, you can replace R5 with a trimmer potentiometer in series with a 10 kΩ resistor. A 100 kΩ trimmer is suitable.

A schematic diagram of the Dual Mini Timer is shown in Fig. 4.14.

Parts List: Dual Mini Timer

IC1	4093 CMOS integrated circuit
X1	Piezoelectric transducer or crystal earphone, Radio Shack 279-073 or equivalent
S1	SPST toggle or slide switch
S2	SPST momentary switch
B1	6 or 9 V (four AA cells or battery)
R1, R3	1,000,000 Ω to 4,700,000 Ω potentiometers or trimmer potentiometers
R2, R4	100,000 Ω, 1/4 W, 5% resistors
R5	47,000 Ω, 1/4 W, 5% resistor
C1, C2	10 µF to 1,000 µF, 12 WVDC electrolytic capacitors
C3	0.022 µF ceramic or metal film capacitor
C4	100 µF, 12 WVDC electrolytic capacitor

Special care must be taken with the positions of the polarized components (electrolytic capacitors and battery). With 1,000 µF capacitors for C1 and C2 and 4.7 MΩ potentiometers for R1 and R3, you can get time delays up to an hour and a half. Operation is as the same with the other timers described in this book.

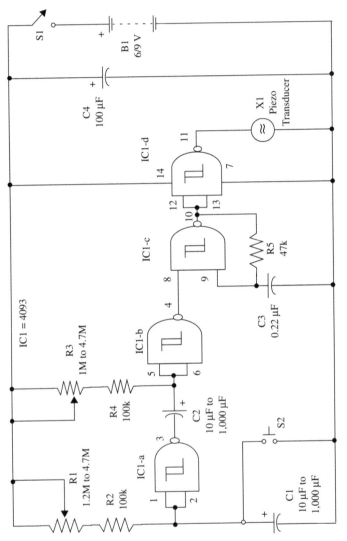

Figure 4.14   Dual Mini Timer.

Project 68: Dual Pulsed-Tone Timer (E) (P)

The first time delay adjustment to this circuit will produce an intermittent or pulsed tone during a second time period, which is adjusted by a second potentiometer. The total time delay can reach an hour and half. The circuit can be used as a darkroom timer, egg timer, or chemical processes timer, and in many other applications.

The timer is powered from four AA cells or a 9 V battery. Current drain is very low (only few milliamperes), which extends the power supply life to many weeks.

A schematic diagram of the Dual Pulsed-Tone Timer is shown in Fig. 4.15.

Parts List: Dual Pulsed-Tone Timer

IC1	4093 CMOS integrated circuit
X1	Piezoelectric transducer or crystal earphone, Radio Shack 279-073 or equivalent
S1	SPST toggle or slide switch
B1	6 or 9 V (four AA cells or battery)
R1, R3	100,000 Ω, 1/4 W, 5% resistors
R2, R4	1,000,000 Ω to 4,700,000 Ω potentiometers or trimmer potentiometers
R5	2,200,000 Ω, 1/4 W, 5% resistor
R6	47,000 Ω, 1/4 W, 5% resistor
C1, C2	10 μF to 1,000 μF, 12 WVDC electrolytic capacitors
C3	0.22μF or 0.47 μF ceramic or metal film capacitor
C4	0.022 μF ceramic or metal film capacitor
C5	100 μF, 12 WVDC electrolytic capacitor

The proper positions of the polarized components (electrolytic capacitors and battery) must be observed. To obtain a compact version, you can use a small printed circuit board.

Time delays are determined by capacitors C1 and C2. With 1,000 μF capacitors, the maximum time delay for each is 45 minutes. Total time delay is more than 90 minutes.

The tone is determined by R6 and C4, and the pulse rate is given by R5 and C3. These components can be changed to alter the final tone.

In operation, adjust the time delays via R2 and R4, and close power switch (S1). After the time delay adjusted by R2, a tone will be produced during a time interval given by the R4 adjustment.

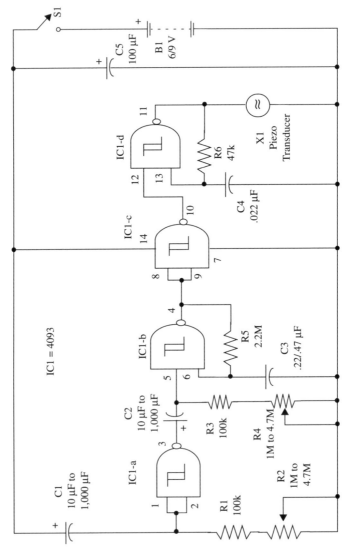

Figure 4.15 Dual Pulsed-Tone Timer.

Project 69: Four-LED Bargraph Timer (E)

This interesting experimental project can be used as an egg timer, darkroom timer, and as a part of other projects. The circuit has four LEDs that turn on sequentially according to the adjustable time delay. The circuit can be set to time delays between a few seconds to one or two minutes, depending on the value of capacitor C1.

The four gates of the 4093 are connected as voltage comparators. A resistor network wired to the input of the gates acts as a voltage divider, determining the moment in which each output goes low and turns on the corresponding LED.

Different values of resistors in the voltage divider are used to obtain a linear response as the capacitor charges through the timer resistors according an exponential characteristic.

A schematic diagram of the Four-LED Bargraph Timer is shown in Fig. 4.16.

Parts List: Four-LED Bargraph Timer

IC1	4093 CMOS integrated circuit
LED1 to LED4	Red common LEDs
R1	100,000 Ω potentiometer
R2	10,000 Ω, 1/4 W, 5% resistor
R3	47,000 Ω, 1/4 W, 5% resistor
R4	33,000 Ω, 1/4 W, 5% resistor
R5	22,000 Ω, 1/4 W, 5% resistor
R6	330,000 Ω, 1/4 W, 5% resistor
R7 to R10	1,000 Ω, 1/4 W, 5% resistor
C1	100 µF to 1,000 µF, 12 WVDC electrolytic capacitor
C2	100 µF, 12 WVDC electrolytic capacitor

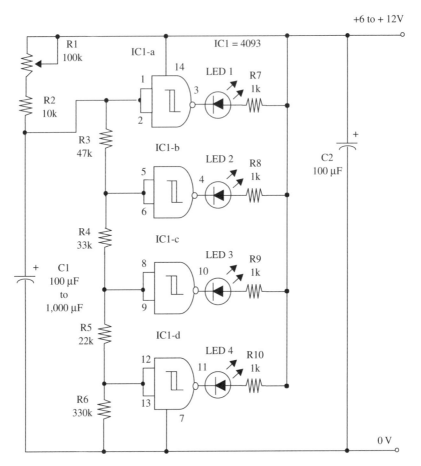

Figure 4.16 Four-LED Bargraph Timer.

Project 70: Bargraph and Relay Timer (E) (P)

In this project, the four LEDs will turn on sequentially. When the last LED is turned on, the relay is energized. You can use this interesting timer to control external loads with a bargraph monitor.

Time delays can be adjusted in a wide range of values, depending on the capacitor used. Time delays between a few seconds and a few minutes are easy to obtain. A mini DPDT relay (Radio Shack 279-079) controls external loads up to 100 W.

Heavy-duty relays can replace the original one, but you must also replace the output transistor with a Darlington. Currents up to 1 A can be controlled by this transistor using a 12 V power supply.

A schematic diagram of this timer is shown in Fig. 4.17.

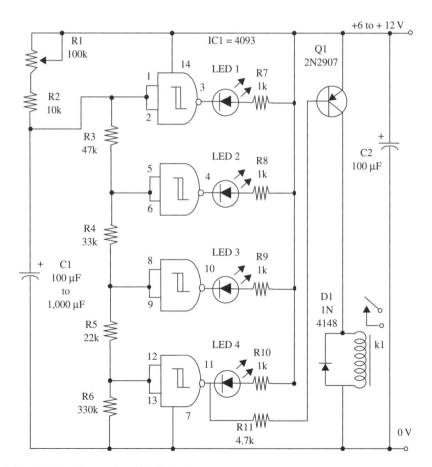

Figure 4.17   Bargraph and Relay Timer.

Parts List: Bargraph and Relay Timer

IC1	4093 CMOS integrated circuit
Q1	2N2907 NPN general purpose silicon transistor
D1	1N4148 general purpose silicon diode
LED1–LED4	Red common LEDs
K1	6 or 12 V relay, up to 100 mA (Radio Shack 275-249, 12 V, 43 mA, is a suitable unit)
R1	100,000 Ω potentiometer
R2	10,000 Ω, 1/4 W, 5% resistor
R3	47,000 Ω, 1/4 W, 5% resistor
R4	33,000 Ω, 1/4 W, 5% resistor
R5	22,000 Ω, 1/4 W, 5% resistor
R6	330,000 Ω, 1/4 W, 5% resistor
R7 to R10	1,000 Ω, 1/4 W, 5% resistors
R11	4,700 Ω, 1/4 W, 5% resistor
C1	100 µF to 1,000 Ω, 12 WVDC electrolytic capacitor
C2	100 µF, 12 WVDC electrolytic capacitor

Positioning of the polarized components (LEDs, electrolytic capacitors, and diodes) must be correct. Note that a small SPST relay can be mounted directly on the solderless board or printed circuit board. Other types of relays can also be used, but according to the case, they should not be mounted on a solderless board. Heavy-duty relays, with 100 to 500 mA coils, can be used by replacing driver transistor Q1 with a PNP Darlington transistor and using a 12 V power supply.

Project 71: Incandescent Lamp Timer (P)

Here we describe a turn-off circuit that gives delays of up to 30 minutes. It can be used with ac incandescent lamps. You can install the device in conjunction with your front door lamp to help you to find the keyhole, in a darkroom, and in many other applications.

The recommended SCR is rated to 4 A but, for a safe operation, do not use the control with greater than 200 W loads (incandescent lamps or heaters—do not use the device with inductive loads or fluorescent lamps).

The circuit works as follows. When S1 is closed, the power supply is on, and C2 begins to charge via R1 and R2. At the same time, a high level is applied to the IC1-a input, giving a low output in this gate and, consequently, a high output level at IC1-b, c, and d outputs. The high output level triggers on the SCR, powering up the incandescent lamp.

When V_p is reached at the IC1-a input, its output goes high and, at the same time, IC1-b, c, and d outputs go low, turning off the SCR and also the incandescent lamp.

To set a new time delay, you have only to press C2 and discharge C2. Time delays of up to 30 minutes can be obtained with a 1,000 µF capacitor and a 4.7 MΩ potentiometer. To reduce the time delays, you can use a low-value capacitor for C2. Observe that the transformer is on when S1 is closed, even when the lamp is off, but power consumption is low in that condition.

A schematic diagram of the Incandescent Lamp Timer is shown in Fig. 4.18.

The SCR must be mounted on a heatsink. To adjust time delays, R1 can be replaced by a trimmer potentiometer.

Positions of the polarized components (diodes, electrolytic capacitors, and SCR) must be correct. All the components can be housed in a plastic box, except the lamp.

Remember that this circuit is in direct contact with the ac power line. Wiring should be carefully checked to make sure that metal parts do not contact wires or components to the ac line. Do not use a metal box to house the components.

In use, first adjust R1 to the desired time delay and close S1. The lamp will glow during the adjusted time delay. To set a new time delay, press S2.

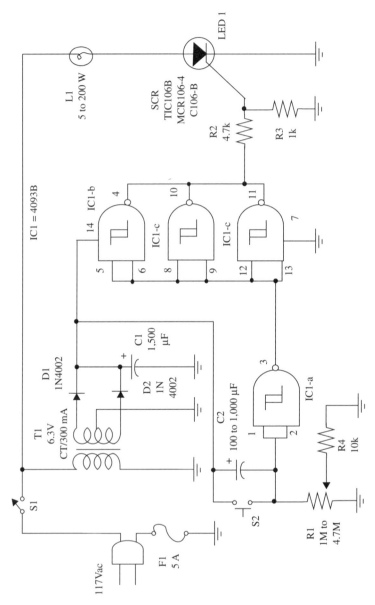

Figure 4.18 Incandescent Lamp Timer

Parts List: Incandescent Lamp Timer

IC1	4093 CMOS integrated circuit
SCR1	TIC106-B or equivalent silicon thyristor (200 V, 4 A)
D1, D2	1N4002 or equivalent 50 V, 1 A silicon diodes
T1	6/3 V, 300 mA to 500 mA secondary, 117 Vac primary, CT transformer or equivalent
S1	SPST toggle or slide switch
S2	SPST momentary switch
F1	5 A fuse and holder
L1	5–200 W incandescent lamp (see text)
R1	1,000,000 Ω to 4,700,000 Ω potentiometer
R2	4,700 Ω, 1/4 W, 5% resistor
R3	1,000 Ω, 1/4 W, 5% resistor
R4	10,000 Ω, 1/4 W, 5% resistor
C1	1,500 µF, 15 WVDC electrolytic capacitor
C2	100 µF to 1,000 µF, 12 WVDC electrolytic capacitor

Project 72: Incandescent Lamp Dual Timer (P)

This circuit is powered from batteries and the ac line simultaneously, driving an incandescent lamp up to 200 W. The circuit turns on an incandescent lamp after an adjustable time delay and automatically turns off the lamp after a second delay. Time delays can be adjusted from seconds to more than half an hour.

The circuit works as follows. When S1 is closed, capacitor C2 charges through R1 and R2 until V_p is reached. At this moment, the IC1-a output goes to a low logic level, and C3 begins to charge through R3 and R4 until V_n is reached.

During the first capacitor charge (C2), IC1-b, c, and d outputs remain at the low level, and the SCR is off. During the second time delay, when C3 is charging, IC1-b, c, and d outputs remain high, and the SCR is triggered on, supplying the lamp from the ac power line.

As soon as the capacitor C3 is charged to V_n, IC1-b, c, and d outputs go to a low logic level again, and this causes the SCR to trigger off. To set a new time delay, you have to discharge C2 by pressing S2.

Remember that the SCR is a half-wave control to ac loads. A diode bridge can be used to give you a full-wave control. Do not use inductive loads or fluorescent lamps in this circuit.

A schematic diagram for the Incandescent Lamp Dual Timer is shown in Fig. 4.19.

Parts List: Incandescent Lamp Dual Timer

IC1	4093 CMOS integrated circuit
SCR	TIC106-B or equivalent silicon thyristor
S1	SPST toggle or slide switch
S2	SPST momentary switch, normally open
R1, R3	1,000,000 Ω to 4,700,000 Ω potentiometers
R2, R4	100,000 Ω, 1/4 W, 5% resistors
R5	4,700 Ω, 1/4 W, 5% resistor
R6	1,000 Ω, 1/4 W, 5% resistor
C1	100 µF, 12 WVDC electrolytic capacitor
C2	10 µF to 1,000 µF, 12 WVDC electrolytic capacitor
F1	5 A fuse and holder
B1	6 V or 9 V (four AA cells or battery)

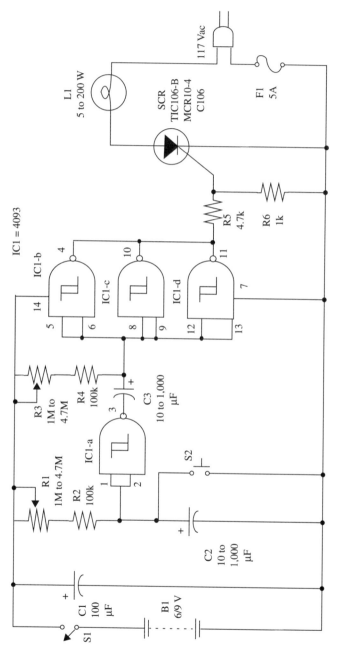

Figure 4.19 Incandescent Lamp Dual Timer.

The positions of the polarized components [power supply (dc) and electrolytic capacitors] must be correct. The SCR must be mounted on a heatsink.

With a 1,000 µF capacitor for C2 and C3, and 4,700,000 Ω potentiometers, the total time delay is up to 90 minutes. Current drain from the dc power supply is very low, extending battery life to several weeks.

Project 73: Automatic Turn-On and Turn-Off Timer (P)

This circuit can be used to turn off a front door lamp when you leave your home and at the same time to activate the alarm system. The circuit is powered from the ac power line or an emergency battery.

To set a new time delay and turn a front door light on during the set time, you need to use a small hole through which a secret push-button can be pressed. Alternatively, you can also replace the push-button with a reed switch that is activated by a small magnet attached to your keyholder as suggested in Fig. 4.20. The circuit can be used with common alarms and powered from the ac power line or battery (the same one that powers the alarm).

Operation is as follows. When S1 is closed, power is on, and C2 begins to charge through R1 and R2. R1 adjusts the time delay. As soon a V_p is reached in pin 2 (IC1-a), its output goes high, and IC1-b, c, and d outputs go low, turning off the relay. During C2 charging, the relay is on, and the lamp glows. After the programmed time delay, the relay turns off, and the alarm's power comes on.

To set a new time delay (when you come back to home), S2 should be pressed. Then, C2 begins a new charge, and the alarm's power supply is cut off during the adjusted time delay.

A schematic diagram of the timer is given in Fig. 4.21.

Proper positioning of the polarized components (diodes, electrolytic capacitors, and transistor) must be observed.

In the basic project, we used a mini DPDT relay (12 V, 43 mA, Radio Shack 273-1365) that could be mounted on a solderless board or universal printed circuit board. If you're using another type of relay, depending on the terminal positions, the layout must be altered.

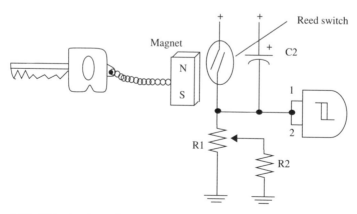

Figure 4.20 This circuit can be activated by a small magnet.

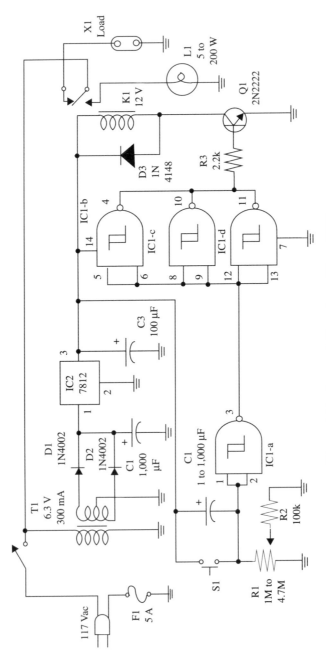

Figure 4.21 Automatic Turn-On and Turn-Off Timer.

Parts List: Automatic Turn-On and Turn-Off Timer

IC1	4093 CMOS integrated circuit
IC2	7812 voltage regulator IC
Q1	2N2222 NPN general purpose silicon transistor
D1, D2, D3	1N4002 silicon rectifiers (50 V, 1 A)
K1	12 V, 100 mA relay, Radio Shack 275-249 or equivalent
T1	Transformer: 12.6 V, 450 mA secondary CT, Radio Shack 273-1365 or equivalent, primary 117 Vac
D4	1N4002 or equivalent silicon rectifier (see text)
S1	SPST toggle or slide switch
S2	SPST reed switch or momentary switch
F1	5 A fuse and holder
R1	1,000,000 Ω to 4,700,000 Ω potentiometer
R2	100,000 Ω, 1/4 W, 5% resistor
R3	2,200 Ω, 1/4 W, 5% resistor
C1	1,000 µF, 25 WVDC electrolytic capacitor
C2	100 µF to 1,000 µF, 16 WVDC electrolytic capacitor
C3	100 µF, 16 WVDC electrolytic capacitor

Heavy-duty relays also can be used, but if they have coils rated in the range of 100 mA to 500 mA, transistor Q1 must be replaced with a Darlington or an NPN power transistor such as the TIP31 or TIP110. The external battery should be connected to the power supply via D4 for ac line failure.

S2 should be installeD in a secret place. If you're using a reed switch as S2, it can be installed under a thin plastic panel to facilitate the magnet's action.

5

Bistable Circuits

Bistable circuits are built by cross-coupling inverting gates with direct wire connections. They are used for data storage, memories, latches, counters, and many other basic configurations. Starting from a basic configuration using two of the four 4093 gates, we present the reader with several interesting projects. As in the other parts of this book, the projects can be used as part of more complex ones or as complete, stand-alone devices.

We also suggest that the reader perform experiments with the components to improve performance for a particular task or to address applications not mentioned in the text.

Project 74: Touch-Activated Bistable (E)

This project shows how a 4093 can be used to drive two LEDs as a sensitive, touch-activated switch. Two of the 4093's four gates are connected as a set-reset flip-flop, with the other two gates as inverters driving the LEDs.

When we touch X1, LED1 turns on, and LED2 turns off. Then, if we touch X2, LED1 turns off, and LED1 turns on again. The very high input resistance of the bistable given by R1 and R2 provides exceptional sensitivity to the circuit.

There are several applications for this experimental circuit. You can use it as a light warning, to demonstrate the bistable action of a 4093, or as part of other projects, replacing LED1 and/or LED2 with powerful transistorized stages driving relays, lamps, or other loads.

The circuit is powered from 6 or 9 volt batteries or power supplies, and current drain is about 20 mA (9 V).

A schematic diagram of the Touch-Activated Bistable is given in Fig. 5.1. Positions of the polarized components (LEDs, electrolytic capacitor) must be observed.

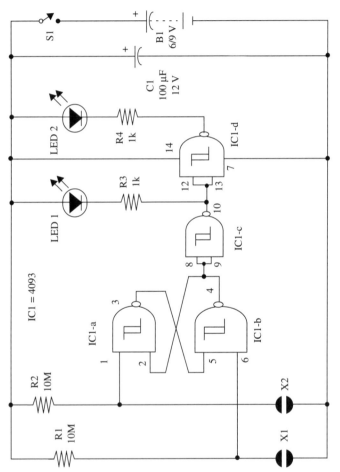

Figure 5.1 Touch-Activated Bistable.

Parts List: Touch-Activated Bistable

IC1	4093 CMOS integrated circuit
LED1, LED2	Common red LEDs, Radio Shack 276-1622 or equivalent
S1	SPST toggle or slide switch
X1, X2	Touch sensors (see text)
R1, R2	10,000,000 Ω, 1/4 W, 5% resistors
R3, R4	1,000 Ω, 1/4 W, 5% resistors
C1	100 μF, 12 WVDC electrolytic capacitor
B1	6 V or 9 V (four AA cells or battery)

X1 and X2 are formed by two metal plates placed in close proximity. They must be touched simultaneously by the fingers to activate the circuit.

If you want to drive powerful light sources such as lamps, or more than two LEDs, you can use transistorized output stages. An NPN general purpose transistor such as the 2N2222 can be driven with a 1 to 4.7 kΩ base resistor and drain of about 100 mA in the collector.

To operate, turn on the power supply (S1) and then touch X1 or X2 to turn the LEDs on or off.

Project 75: Touch Turn-On/Off Relay (E) (P)

Lamps, small home appliances, tools, and other ac-powered devices can be controlled by the touch of your fingers with this simple circuit. You can turn on any load by touching sensor X1 and turn it off by touching X2. There is no shock hazard, as the control is completely isolated from the ac power line.

The control is powered from four or six AA cells or, if you prefer, a power supply ranging from 6 to 12 V, depending on the relay coil. *Don't use transformerless power supplies: they are not isolated from the ac power line and can cause severe shocks.*

A schematic diagram of the Touch Turn-On/Off Relay is given in Fig. 5.2.

Parts List: Touch Turn-On/Off Relay

IC1	4093 CMOS integrated circuit
Q1	2N2222 NPN general purpose silicon transistor
D1	1N4148 or equivalent general purpose silicon diode
K1	6 V or 12 V mini DPDT or SPST relay (see text), Radio Shack 275-2488
R1, R2	10,000,000 Ω, 1/4 W, 5% resistor
R3	2,200 Ω, 1/4 W, 5% resistor
C1	100 μF, 16 WVDC electrolytic capacitor
X1, X2	Sensors (see text)

The relay wiring is determined by the type of relay you intend to use. You can use a mini 1 A DPDT (Radio Shack 275-249) 12 V, 280 Ω, 43 mA, and wire it as shown in the figure, or use other types of relays according the load requirements. A 10 A SPDT mini relay (Radio Shack 275-278) is suitable for heavy-duty appliances. Of course, the power-supply voltage should be the same as the relay coil voltage.

As a simple rule, you can use 6 or 12 V relays with coil currents ranging from 10 to 100 mA and contacts up to 10 A, or rated according the task you have in mind.

Proper positioning of the polarized components (diode D1, electrolytic capacitor, and the transistor) must be observed. Sensors X1 and X2 are made as we described in Project 74.

Don't use a metallic box to house the device, as there are parts connected directly to the ac power line. Be sure that there is no power line contact with the low-voltage circuit to avoid shocks and dangerous short circuits.

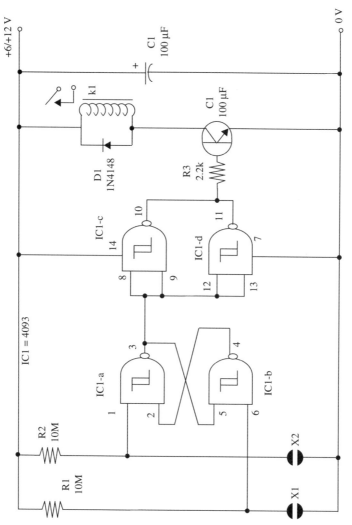

Figure 5.2 Touch Turn-On/Off Relay.

The load is connected as shown for other projects that use relays in an "on" state when the relay coil is energized. However, you can also use the NC (normally closed) contacts to turn off a load when the relay is energized.

Remember that current requirements are high when the relay is energized and low when the transistor is off (coil not energized). That is an important factor to consider if you are using batteries to power the unit.

Project 76: Touch-Controlled Motor (E) (P)

You can control a small electric dc motor, with your fingers determining whether it runs forward or backward. When you touch sensor X1, the motor runs forward, and backward when you touch X2. Small dc motors can be controlled by this circuit, since the current requirements are in the relay's contact range. You can use the circuit in model railroads, small robots and toys, remote control, and other projects.

The circuit can be powered from 6 V to 12 V power supplies according the motor's requirements, or you can use a separate power supply for the motor.

Operation can be described as follows. When you touch X1, IC1-b input (pin 6) goes low. At the same time, its output (pin 4) goes high. This makes the IC1-a output (pin 3) change its state to low, giving a stable state to the flip-flop formed by the two gates.

The low state of the IC1-a output determines a high level at the transistor base, so the relay is energized. The motor is powered in the direct mode and runs forward.

Now, if you touch sensor X2, IC1-a output (pin 3) goes to a high level. That dictates a low level at IC1-c and d outputs, and the transistor Q1 is off. Therefore, the relay is in the non-energized condition, and the motor runs in reverse.

A schematic diagram of the circuit is given in Fig. 5.3.

Parts List: Touch-Controlled Motor

IC1	4093 CMOS integrated circuit
Q1	2N2222 NPN general purpose silicon transistor
D1	1N4148 general purpose silicon diode
X1, X2	Sensors (see text)
K1	6 V or 12 V relay, Radio Shack 275-249 or equivalent (see text)
M	6 V or 12 Vdc motor (up to 1 A)
R1, R2	10,000,000 Ω, 1/4 W, 5% resistors
R3	4,700 Ω, 1/4 W, 5% resistor
C1	100 µF, 16 WVDC electrolytic capacitor

Relay K1 is a DPDT 1 A, 12 V relay (Radio Shack 275-249 or equivalent), but you can also use a 6 V DPDT relay with coil current ranging from 10 to 100 mA. Contacts current are chosen according the motor's requirements.

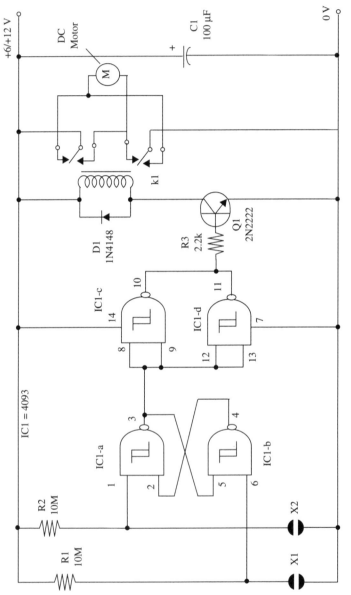

Figure 5.3   Touch-Controlled Motor.

Proper positioning of the polarized components (electrolytic capacitor, transistor, and the diode) must be observed. Sensors are made as described in Project 74.

If you are using this project in a model railroad, you can make an interesting and sensitive sensor with conductive foam glued to the bottom of a boxcar. Contact with the sensor can reverse the train movement. Before installing the unit, be sure that the motor is wired to provide the desired direction of movement.

Project 77: Unilateral Counter (E) (P)

This circuit uses a sensor based on two photocells and has several interesting applications. It can be used as part of a more complex event counter, in scientific experiments, and in other interesting applications. Incoming counts can be totalled by a simple CMOS or TTL counter or a mechanical counter.

The light input produces a square wave in the output only when the light falling on the sensor is cut in a unilateral mode.

Sensors are light dependent resistors (LDRs) or cadmium sulfide (CdS) photocells (Radio Shack 276-1657 or equivalent), and sensitivity can be adjusted to the intended application using R2 and R4. The variable resistors, R1 and R3, should have values appropriate to the light source and the distance from the sensors.

Weak light sources in a dark medium can be detected with large values for all these components. Small values (100 kΩ for the variable resistors and 10 kΩ for the fixed resistor) are used with strong light sources. Output is CMOS compatible, but with an appropriate interface you can drive TTL or other logic families.

A schematic diagram of the Unilateral Counter is shown in Fig. 5.4.

Parts List: Unilateral Counter

IC1	4093 CMOS integrated circuit
LDR1, LDR2	LDRs or CdS photocells, Radio Shack 276-1657 or equivalent
R1, R3	10,000 to 100,000 Ω, 1/4 W, 5% resistors (see text)
R2, R4	100,000 Ω to 1,000,000 Ω potentiometers (see text)
C1	100 μF, 16 WVDC electrolytic capacitor
L1, L2	Common lamps, light sources (see text)

Short cables should be used to wire the sensors. If you need long wires for the intended application, you probably should use shielded cables. The shield is connected to the circuit positive rail.

C1 is the only polarized component. Position this component properly, as shown in the figure. The power supply is in the range between 6 and 12 V. To prevent false inputs with fast cuts in the light source, you can wire 1,000 pF capacitors between pin 1 and 6 to the 0 V reference.

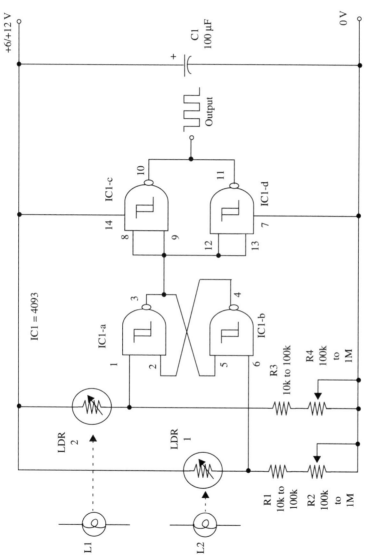

Figure 5.4 Unilateral Counter.

Project 78: Bistable Sonic Relay (E) (P)

Clap your hands, and turn on a lamp; touch the sensor with your fingers, and you'll turn the lamp off. This sonic relay has a bistable action and can be used to control small appliances, tools, lamps, and many other electric devices.

You can also use this project as an alarm to detect any strange sounds in your home and to turn on a siren or other annunciator.

The circuit is powered from a 6 V or a 12 V supply according the relay used. When the relay is not energized, current drain is very low. Sensitivity can be adjusted to a wide range of sound levels by R3 to prevent false operation.

A schematic diagram of the Sonic Relay is given in Fig. 5.5.

Parts List: Bistable Sonic Relay

IC1	4093 CMOS integrated circuit
Q1, Q2	2N2222 NPN general purpose silicon transistors
D1	1N4148 general purpose silicon diode
K1	6 V or 12 V relay, mini 1 A DPDT 12V, Radio Shack 275-249 or equivalent
MIC	Electret mike, Radio Shack 270-090 or equivalent
X1	Touch sensor, as described in Project 74
R1	4,700 Ω (6 V) or 10,000 Ω (12 V), 1/4 W, 5% resistor
R2	2,200,000 Ω, 1/4 W, 5% resistor
R3	1,000,000 Ω potentiometer
R4	22,000 Ω, 1/4 W, 5% resistor
R5	4,700,000 Ω, 1/4 W, 5% resistor
R6	4,700 Ω, 1/4 W, 5% resistor
C1	10 µF, 12 WVDC electrolytic capacitor
C2	100 µF, 16 WVDC electrolytic capacitor

Proper positioning of the polarized components (electret mike, electrolytic capacitors, diode D1, and the transistor) must be observed.

For a remote connection to the mike, use a shielded cable. The shield should be connected to the negative power supply line. Sensor X1 is made as shown in Project 74.

The relay's coil is rated according the power-supply voltage, and contact requirements depend on the load you intend to control. You can

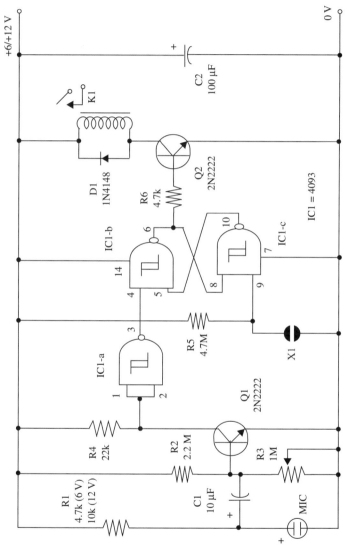

Figure 5.5 Bistable Sonic Relay.

use a mini DPDT 1 A relay for small loads (Radio Shack 275-249) or a heavy-duty relay such as a mini SPDT 10 A for other loads (Radio Shack 275-248 or equivalent).

To adjust the circuit, close R3 and turn on the power supply. Clap your hands and at the same time open R3 until the relay turns on. To reset the circuit, touch sensor X1.

Project 79: Bistable II (E)

All the previous bistable projects we have shown used two sensors or two inputs to set and reset a load. This circuit and the next we show are different: the use only one sensor or one input to set and reset the bistable.

Our experimental Bistable II acts on a relay, closing its contacts when you press S1 the first time and opening the contacts at the second touch.

This circuit can be used to demonstrate the underlying operational principle as used for remote control, alarms, scientific experiments, and many other applications limited only by the reader's imagination.

S1 can be replaced by a reed switch for alarms and applications in which magnetic fields are used to turn the loads on and off. As a load, you can use lamps, small dc motors, and so forth up to 1 A. To drive inductive loads, wire a parallel diode to protect the transistor against high-voltage spikes.

The power supply is chosen to match load requirements. The circuit will operate in the voltage range of 6 V to 12 V without changing any components.

A schematic diagram of the Bistable II is shown in Fig. 5.6.

Parts List: Bistable II

IC1	4093 CMOS integrated circuit
Q1	TIP120 or equivalent Darlington power transistor, NPN
S1	SPST momentary switch
R1	47,000 Ω potentiometer
R2	100,000 Ω, 1/4 W, 5% resistor
R3	2,200 Ω, 1/4 W, 5% resistor
C1	0.47 µF ceramic or metal film capacitor
C2	100 µF, 16 WVDC electrolytic capacitor

To drive loads up to 500 mA, mount transistor Q1 on a heatsink. You can replace the Darlington transistor with a power FET without any other changes to the circuit.

R1 is adjusted to give a set and reset action by pressing S1. In experimental applications, use an incandescent lamp as the load.

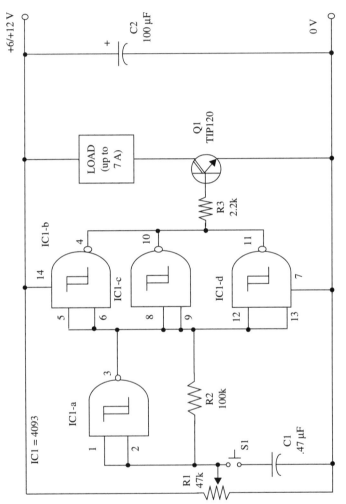

Figure 5.6   Bistable II. S1 sets and resets this circuit.

Project 80: Touch-Activated Relay II (E) (P)

This project uses a new bistable two-gate configuration and only one touch sensor. The circuit can be used in demonstrations, as a component of alarms, in remote control, and for many other applications.

The power supply can range from 6 to 12 V, and current requirements are very low when the relay is off.

Two of the four gates of a 4093 are used as a trigger from a touch sensor. This circuit triggers a flip-flop formed by the other two 4093 gates. A transistor in the flip-flop output drives a relay from the existing logic level.

Sensitivity is determined by resistor R1, and you can vary it in the range between 100 kΩ and 10 MΩ according the application you have in mind. Higher values give greater sensitivity. You can also replace R1 with a 10 MΩ potentiometer to obtain sensitivity control for the circuit.

The power supply choice is determined by the relay. For 12 V applications, you can use a mini 1 A DPDT relay (Radio Shack 275-249), but any relay with 6 V or 12 V coils and a current drain from 10 mA to 100 mA can be used in this project.

A schematic diagram of Touch Activated Relay II is given in Fig. 5.7.

Parts List: Touch-Activated Relay II

IC1	4093 CMOS integrated circuit
Q1	2N2222 NPN general purpose silicon transistor
D1	1N4148 general purpose silicon diode
X1	Touch sensor, as in Project 74
R1, R2, R3	10,000,000 Ω, 1/4 W, 5% resistors
R4	4,700 Ω, 1/4 W, 5% resistor
C1	0.22 µF ceramic or metal film capacitor
C2	100 µF, 16 WVDC electrolytic capacitor
K1	6 V or 12 V relay (see text)

Proper positioning of the polarized components (diode, electrolytic capacitor, transistor) must be observed. Sensor X1 is made as described in Project 74. Short wires are preferable to connect this sensor to the circuit. Longer wires to the sensor should be shielded. Connect the shield to the positive power supply line. Change R1 according the desired sensitivity or replace it with a 10 MΩ potentiometer.

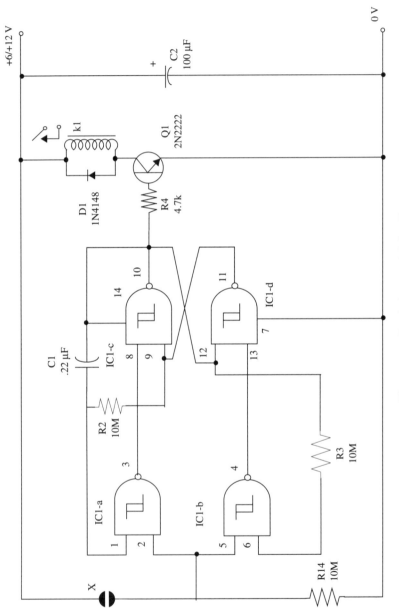

Figure 5.7 Touch-Activated Relay II.

If you want to control heavy-duty loads, use a 10 A mini SPST relay (Radio Shack 275-248). Don't power the circuit from a transformerless power supply, as they don't provide isolation from the ac power line, and this can result in a severe shock hazard when someone touches the sensor.

Project 81: Bistable Light Remote Control (P)

This project shows how to modify Project 80 to obtain a remote control for home appliances and alarms. Replacing the touch sensor (X1) by a photocell (LDR or CdS), we can trigger a relay from light sources such as a flashlight, a match, or a mirror.

The circuit includes a sensitivity control (R2) and can operate with very weak light sources such as a single match several feet from the sensor. To get greater sensitivity, you can install the sensor (LDR) in an opaque cardboard tube and fix a convergent lens in its front end.

The circuit can be powered from 6 V or 12 V supplies according the relay, and current drain is very low when the relay is off. This is very important if you intend to use the project in battery-powered applications.

A schematic diagram of the Bistable Light Remote Control is shown in Fig. 5.8.

Parts List: Bistable Light Remote Control

IC1	4093 CMOS integrated circuit
Q1	2N2222 NPN general purpose silicon transistor
D1	1N4148 general purpose silicon diode
LDR	Cadmium sulfide photocell, Radio Shack 276-1657 or equivalent
K1	6 V or 12 V relay (see text)
R1	4,700 Ω, 1/4 W, 5% resistor
R2	1,000,000 Ω potentiometer
R3, R4	10,000,000 Ω, 1/4 W, 5% resistor
R5	4,700 Ω, 1/4 W, 5% resistor
C1	0.22 µF ceramic or metal film capacitor
C2	100 µF, 16 WVDC electrolytic capacitor

Use a shielded cable if the sensor is remote from the circuit. The relay is chosen according the power supply voltage (or vice versa), and there are several common types that can be used in this project. Relays with 6 V or 12 V coils draining between 10 and 100 mA can be used. Small appliances can be controlled by a 1 A mini DPDT relay, Radio Shack 275-249.

Proper positioning of the polarized components (diode, electrolytic capacitor, and transistor) must be observed.

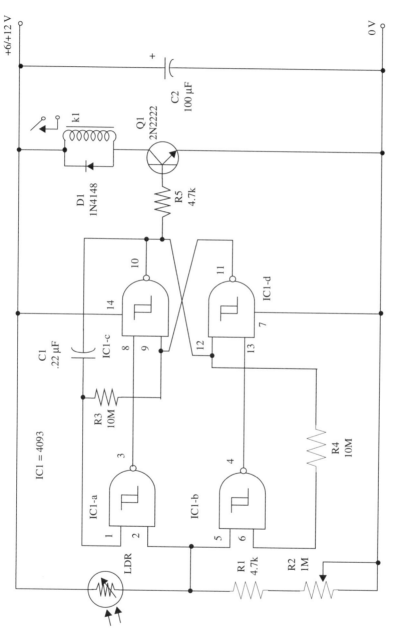

Figure 5.8 Bistable Light Remote Control.

Sensitivity is adjusted by R2. Set it according the contrast between the control source and the ambient light. If you want a directional operation, install the LDR into an opaque cardboard tube and fix a convergent lens in the front end of it. The position of the lens is determined by its focal distance; the focus should be on the LDR surface.

Project 82: Coin Tosser (P)

This circuit simulates the flipping of a coil merely by pushing S1. Of course, the electronic version, if used for important decisions, can't be loaded or weighted—it is 100 percent random.

The circuit has two LEDs that flick alternatively when you push S1. Afterward, the circuit doesn't stop immediately but continues during a time period provided by R1 and C1. After this time, only one LED will be on, and that determines the winner.

A schematic diagram of the Coin Tosser is shown in Fig. 5.9.

Parts List: Coin Tosser

IC1	4093 CMOS integrated circuit
LED1, LED2	Red and green LEDs
S1	SPST momentary switch
S2	SPST slide or toggle switch
B1	6 V or 9 V, four AA cells or battery
R1, R2, R3	1,000,000 Ω, 1/4 W, 5% resistors
R4, R5	1,000 Ω, 1/4 W, 5% resistors
C1	2.2 µF, 16 WVDC electrolytic capacitor
C2	0.22 µF ceramic or metal film capacitor
C3	0.022 µF ceramic or metal film capacitor
C4	100 µF, 16 WVDC electrolytic capacitor

Proper positioning of the polarized components (LEDs, electrolytic capacitors, and power supply) must be observed.

You can change C1, C2, and C3 to vary the final performance of the Coin Tosser.

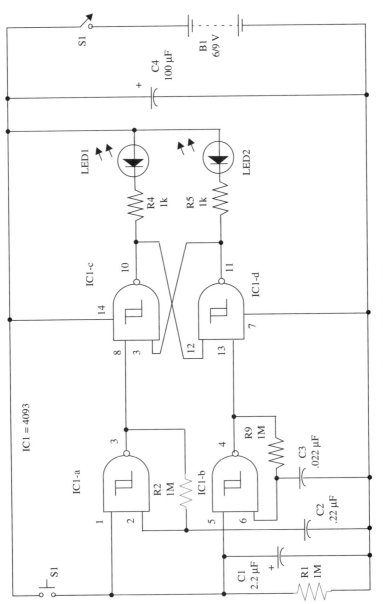

Figure 5.9 Coin Tosser.

6

Alarms

The low current drain of CMOS integrated circuits makes them ideal for use in a variety of electronic alarm projects. Several of this projects are described in this chapter, again using the 4093. These projects include both simple and advanced burglar alarms, steam and water alarms, car alarms, over- and under-temperature alarms, and many others.

As in the previous chapters, each the projects can be used alone, as a complete device, or as part of a more complex project. Use your imagination to put all these circuits together.

Project 83: Swimming Pool or Rain Alarm (P)

A sensor fixed on the side of a swimming pool can detect a wave formed if someone falls into the water. Another kind of sensor can be used in the same circuit to detect rain or water. The basic version drives a small piezoelectric transducer, but you can get more audio from the circuit by changing the output stage to one of the transistorized configurations shown earlier in this book.

Reset is by touch. You only need to touch two plates, one near the other, to stop the alarm.

Power comes from AA cells or from a 9 V battery. When the sound is off, the current drain is very low, extending the battery life to several weeks.

Sensor wiring can be long, which is important if you want to install the transducer some distance from the swimming pool. If the distance is more than 30 feet, we recommend shielded cable for this connection.

A schematic diagram of the Swimming Pool Alarm is given in Fig. 6.1.

For a swimming pool alarm, sensor X1 is constructed with two common wires as shown in the figure. The wires are separated by a distance of 1 to 2 inches and stand 1/2 to 1 inch above the water.

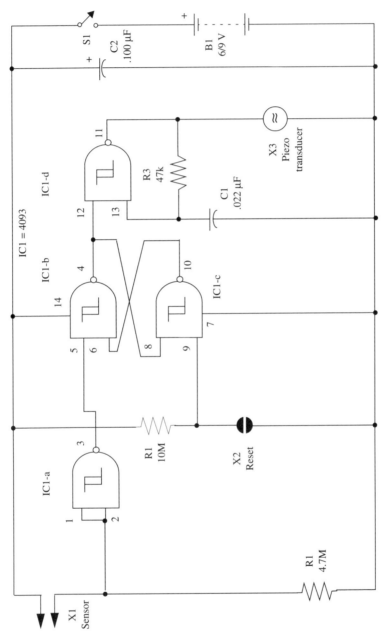

Figure 6.1 Swimming Pool or Rain Alarm.

Parts List: Swimming Pool or Rain Alarm

IC1	4093 CMOS integrated circuit
X1	Sensor (see text)
X3	Piezoelectric transducer or crystal earpiece, Radio Shack 273-073 or equivalent.
X2	Touch sensor (see text)
S1	SPST toggle or slide switch
B1	6 V or 9 V, four AA cells or battery
R1	4,700,000 Ω, 1/4 W, 5% resistor
R2	10,000,000 Ω, 1/4 W, 5% resistor
R3	47,000 Ω, 1/4 W, 5% resistor
C1	0.022 μF ceramic or metal film capacitor
C2	100 μF, 16 WVDC electrolytic capacitor

For the rain or water alarm, a sensor is formed by two metal (copper or aluminum) screens separated by a piece of porous paper or tissue with some salt. Remember that, in this version, to reset the alarm, you have to replace the piece of paper or tissue with a dry one before acting on X2.

Correct positioning of the polarized components must be observed. The reset sensor is made as described in Project 74.

Project 84: Pendulum Intermittent Alarm (E) (P)

Any movement that swings the pendulum sensor triggers this alarm, turning a relay on and off. The on and off rate of the relay is determined by components that can be varied.

The circuit can be used to protect cars and other large objects, the home, and other items, as the sensor can be used to detect any kind of movement. You can also use this circuit to detect movement in scientific experiments.

The relay can drive powerful warning systems such as sirens, horns, lamps, and so on. The circuit can be powered from common cells or batteries, and current drain is very low (about 5 mA) when the relay is off.

A schematic diagram of the Pendulum Intermittent Alarm is given in Fig. 6.2.

Parts List: Pendulum Intermittent Alarm

IC1	4093 CMOS integrated circuit
Q1	2N2222 NPN general purpose silicon transistor
D1	1N4148 general purpose silicon diode
K1	6 V or 12 V relay (see text)
X1	Pendulum sensor (see text)
X2	Touch sensor or momentary SPST switch
R1	1,000,000 Ω, 1/4 W, 5% resistor
R2	10,000,000 Ω, 1/4 W, 5% resistor
R3	2,200,000 Ω to 4,700,000 Ω, 1/4 W, 5% resistor (see text)
R4	4,700,000 Ω, 1/4 W, 5% resistor
C1	0.22 μF or 0.47 μF metal film or ceramic capacitor (see text)
C2	100 μF, 16 WVDC electrolytic capacitor

The relay coil is chosen to match the power supply voltage. You can use a mini DPDT, 1 A relay, Radio Shack 275-249 or, for heavy loads, a 10 A mini SPDT relay, Radio Shack 275-248. These relays are rated to 12 V and drain only 38 mA (coil resistance = 320 Ω).

The turn-on and turn-off frequency is determined by R3 and C1. You can vary the values of these components to alter the intermission rate. Values can range as shown in the schematic diagram.

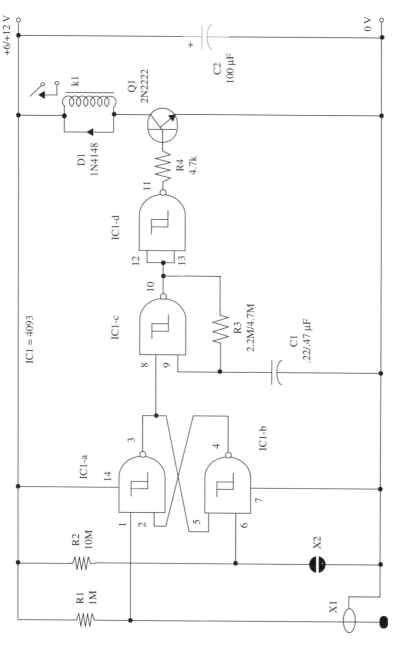

Figure 6.2 Pendulum Intermittent Alarm.

Reset is made by a touch switch as described in Project 83, but you can also replace this sensor with a common SPST momentary switch. Proper positioning of the polarized components (diode and electrolytic capacitor) should be observed.

The sensor's details are shown in Fig. 6.3. Note that operation occurs when the vertical bare wire touches the ring due any vibration.

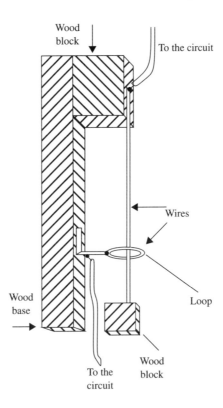

Figure 6.3   The sensor is made as shown above.

Project 85: General Purpose Remote Alarm (P)

This is a remotely operated alarm that can be used to protect your home or other locations. The circuit is a non-latching alarm that uses two kinds of sensors. The General Purpose Alarm operates when any of the input sensors (X1, X2, and X3) are open or when the magnetic intermediate sensors (X4 and X5) is closed.

Dozens of sensors can be wired in parallel or series, and each will cause the alarm to operate when activated. Current in the sensors is only few milliamperes, so they can be placed hundreds of feet away from the circuit without risk of problems caused by high cable resistance. The circuit drives a relay that can be used to control sound sources such as sirens, horns, and so forth.

Power comes from 6 V or 12 V batteries, depending on the relay. Standby current is only 2 mA when the alarm is in the off condition, so it causes a negligible drain on the supply battery.

The sensors can be microswitches or reed types, and the alarm can be made to operate whenever a door or window is opened or when an object moves beyond a preset limit. You can also use pressure-pad switches.

Note that the alarm operates when X1, X2, or X3 is open, or when X4 or X5 are closed, so you can use both normally open or normally closed switches.

A schematic diagram of the General Purpose Remote Alarm is shown in Fig. 6.4.

Parts List: General Purpose Remote Alarm

IC1	4093 CMOS integrated circuit
Q1	2N2222 NPN general purpose silicon transistor
D1, D2, D3	1N4148 general purpose silicon diodes
K1	6 V or 12 V relay (see text)
X1, X2, X3	Normally closed sensors (see text)
X5, X6	Normally open sensors (see text)
R1, R2	1,000,000 Ω, 1/4 W, 5% resistors
R3	100,000 Ω, 1/4 W, 5% resistor
R4	4,700 Ω, 1/4 W, 5% resistor
C1	100 µF, 16 WVDC electrolytic capacitor

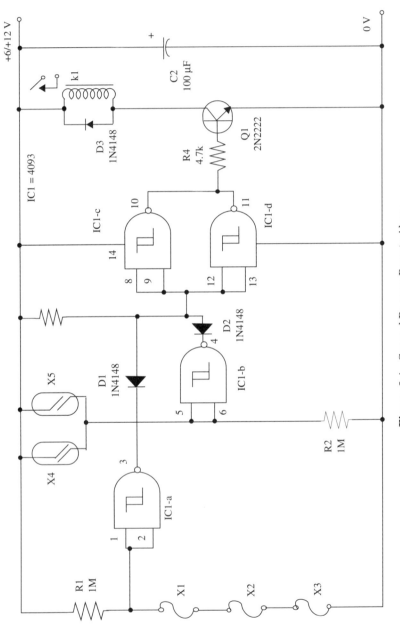

Figure 6.4 General Purpose Remote Alarm.

A DPDT mini relay rated at 1 A (Radio Shack 275-249) can be used with a 12 V power supply and loads up to 1 A. For more powerful loads, you can use a 10 A SPDT mini relay, Radio Shack 275-248.

To reset the alarm you have to close any open sensor (X1, X2, or X3) or open any closed sensor (X4 or X5). Latching alarms are given in other projects within this book.

Project 86: Bistable Light Alarm (P) (E)

You can use this alarm in several interesting applications in the home or in industry. It can be used to sound an alarm when light enters a normally dark room or when a flashlight shines on the sensor. Our circuit has its own sound source, a piezoelectric transducer, but you can get more sound using any of our other power stages.

The circuit is versatile and will work with almost any light-dependent resistor (LDR) or cadmium sulfide photocell such as the Radio Shack 276-1657, with face diameters ranging from 1/8 to 1 inch.

The LDR and R1 form a potential divider that supplies gate drive to IC1-a. When the circuit turns on, IC1-d is activated as an oscillator until the flip-flop formed by IC1-b and c is reset.

IC1-d acts as an audio oscillator, driving a piezoelectric transducer. If you need a more powerful audio output, you can use a power output stage with one or two transistors driving a loudspeaker, as shown in many other projects in this book.

A schematic diagram of the Bistable Light Alarm is shown in Fig. 6.5.

Parts List: Bistable Light Alarm

IC1	4093 CMOS integrated circuit
X1	LDR or CdS photocell, Radio Shack 276-1657 (see text)
X2	Piezoelectric transducer, Radio Shack 273-073 or equivalent
X3	Touch sensor or momentary SPST switch
S1	SPST slide or toggle switch
B1	6 V or 9 V, four AA cells or battery
R1	1,000,000 Ω potentiometer or trimmer potentiometer
R2	10,000,000 Ω, 1/4 W, 5% resistor
R3	47,000 Ω, 1/4 W, 5% resistor
C1	0.022 µF ceramic or metal film capacitor
C2	100 µF, 12 WVDC electrolytic capacitor

Transducer X1 is a piezo element such as the Radio Shack 273-073 or equivalent. You can also use as transducer a crystal earphone, but these elements will give less audio output than the original.

R1 adjusts sensitivity, and you can both use a potentiometer or trimmer pot to fit the intended application. Sensitivity is so high that the

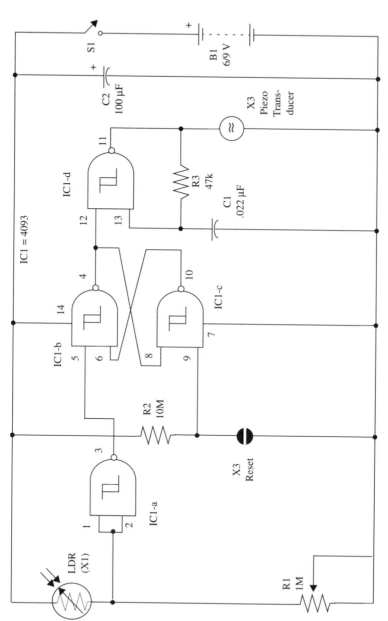

Figure 6.5 Bistable Light Alarm.

device can be turned on with light levels too small to be detected by the human eye.

Directional action is obtained by mounting the LDR into a opaque cardboard tube with a convergent lens. The LDR's sensitive surface should be near the lens' focus. Sensitivity is also proportional to the lens diameter.

Changing this device into a dark-activated alarm is very easy: you simply transpose the LDR and R1 positions. The frequency of the audio signal can be altered by changing both C1 and R3.

Project 87: Freezer Alarm (P)

It is very important that doors of freezers (and refrigerators) remain normally closed. The alarm we describe in this project will go off when the door is opened.

The project is based on a light dependent resistor (LDR) or cadmium sulfide (CdS) photocell and works as follows. As soon as the door of a freezer is opened, light falls onto the LDR, activating the circuit (i.e., IC1-a output goes from low to high). The circuit is then actuated, and a warning tone is produced by a piezoelectric transducer until the door is closed again (the circuit is not timed).

IC1-b acts as a low-frequency oscillator, and IC1-c as an audio oscillator. The oscillators' signals are combined by IC1-d to drive the piezoelectric transducer with an intermittent tone. The circuit can be powered from AA cells or a 9 V battery, and current drain is on the order of 0.5 mA when off and about 5 mA when the tone is on.

A schematic diagram of the Freezer Alarm is given in Fig. 6.6.

Parts List: Freezer Alarm

IC1	4093 CMOS integrated circuit
X1	Piezoelectric transducer or crystal earpiece, Radio Shack 273-073
LDR	CdS photocell (any type), Radio Shack 276-1657 or equivalent
S1	SPST toggle or slide switch
B1	6 V or 9 V, four AA cells or battery
R1	1,000,000 Ω trimmer pot.
R2	2,200,000 Ω, 1/4 W, 5% resistor
R3	47,000 Ω, 1/4 W, 5% resistor
C1	0.22 µF to 0.47 µF ceramic or metal film capacitor
C2	0.022 µF ceramic or metal film capacitor
C3	100 µF, 12 WVDC electrolytic capacitor

Proper positioning of the polarized components (electrolytic capacitor and power supply) should be observed.

You can use any type of LDR in this project. Types with diameters ranging from 1/4 to 1 inch are suitable, as sensitivity can be adjusted by R1. All the components can be housed into a plastic box, but it needs to incorporate a hole to allow the light to fall onto the LDR and generated tone to escape. The sound pulses can be altered by varying R2 or C1. C1 can range from 0.22 µF to 0.47 µF. Audio tone is determined by R3 and C2. These component values can also be varied.

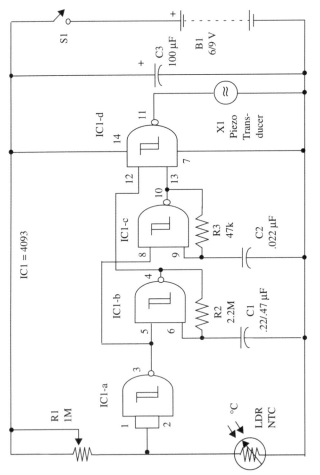

Figure 6.6 Freezer Alarm.

Project 88: Bistable Remote Control (P) (E)

You can turn small appliances on and off using a flashlight as a transmitter with this remote control. Flash the light once, and you turn on the load. Flash it again, and you turn the load off. You can also use this remote control to open your garage door or to turn on a lamp.

The circuit can be powered from 6 V or 12 V power supplies, depending on the relay you have in hand to the project. The sensor cells should not be placed in a position where they are excessively illuminated, and they should be separated from one another by a minimum distance of 10 inches. This separation is needed to keep the flashlight from triggering both cells at the same time.

Loads up to 1 A can be controlled with a mini DPDT 1 A relay (Radio Shack 275-249), and you can also control powerful loads by replacing the original relay with a 10 A SPDT 10 A relay (Radio Shack 275-248).

The sensors are common CdS photocells or LDRs with diameters ranging from 1/4 to 1 inch. Any type is suitable for this project.

The power supply depends on the intended application. You can use AA cells or a battery for a portable use, or a 117 Vac or 12 Vdc power-supply for 500 mA or more.

A schematic diagram of the device is shown in Fig. 6.7.

Parts List: Bistable Remote Control

IC1	4093 CMOS integrated circuit
Q1	2N2222 NPN general purpose silicon transistor
D1	1N4148 general purpose silicon diode
K1	6 V or 12 V relay (see text)
LDR1, LDR2	CdS photocell (see text), Radio Shack 276-1657 or equivalent
R1, R2	1,000,000 Ω potentiometer or trimmer pot
R3	4,700 Ω, 1/4 W, 5% resistor
C1	100 µF, 16 WVDC electrolytic capacitor

Each photocell is housed in an opaque cardboard tube. Directional action is given by a convergent lens placed in front of each LDR.

The device is housed in a plastic box. Dimensions of this box are suggested in Fig. 6.8. The LDRs should be separated by 10 inches or more.

Sensitivity adjustments are made by R1 and R2. You can use common potentiometers or trimmer potentiometers on the board. The load is connected as shown in the figure. Take care with the load's connections to avoid shock hazards, as the circuit is powered from the ac power line.

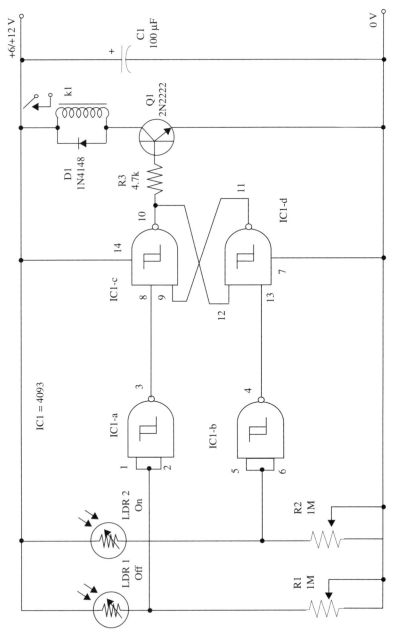

Figure 6.7 Bistable Remote Control.

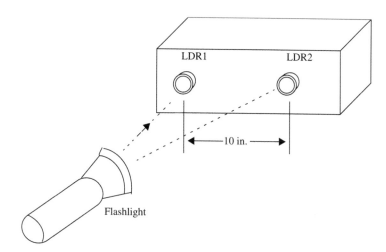

Figure 6.8   The LDRs should be mounted with a separation distance of ≥ 10 inches.

Operation is very easy. Place the remote control and the flashlight in the normal operation position within the normal ambient illumination. Adjust R1 and R2 to turn the relay on and off with short-duration flashes.

Project 89: Delayed Turn-On Alarm I (E) (P)

This circuit will power the alarm on after a time delay of about 20 seconds. This time delay is determined by R1 and C1. You can alter the delay period by changing both R1 and C1. Values such as C1 = 1,000 μF and R1 = 1,000,000 Ω will give time delays of about 15 min.

Normally closed sensors are used in this alarm, and operation occurs when any sensor is opened. You can use wires or reed switches as sensors. Dozens of them can be wired in series, and each one will cause the alarm to operate. Currents of only a few milliamperes pass through sensors' wiring, so they can be placed hundreds of feet away from the circuit without risk of malfunction.

The circuit drives a loudspeaker with about 1 W of audio tone. If you use a 6 V power supply, you can use small general purpose transistors such as the 2N2222 and 2N2907 in the output stage. However, if you power the circuit with 12 V supplies, you should use such powerful transistors as the pair formed by TIP31 and TIP32. In this case, the transistors must be mounted on heatsinks.

The audio tone frequency is determined by R4 and C2. You can vary both within a large range of values. C2 can range from 0.01 to 0.1 μF, and R4 from 22 kΩ to 1 MΩ.

Better audio reproduction is obtained by enclosing the loudspeaker in an enclosure. LED1 is used to indicate whether the alarm is on or off. The LED will glow when the alarm is off. The current drain is very low when the alarm is off.

A schematic diagram of the alarm is shown in Fig. 6.9.

Transistor selection depends on the power supply voltage. With a 12 V supply, you have to use transistors TIP31 and TIP32 in the output stage. S1 is a momentary SPST switch and is used to give the turn-on delay. You have to press this switch when leaving your home. This will give you time to pass through the door without setting off the alarm.

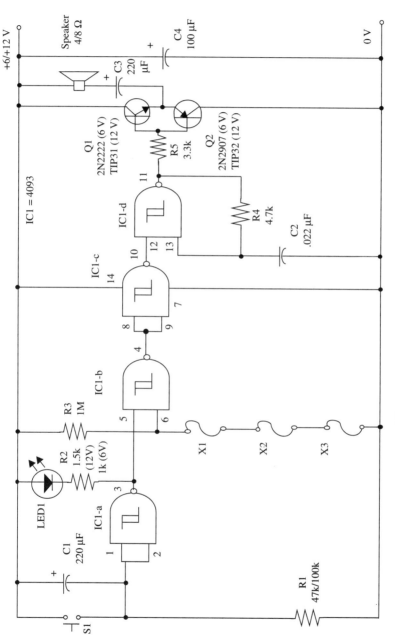

Figure 6.9   Delayed Turn-On Alarm I

Parts List: Delayed Turn-On Alarm I

IC1	4093 CMOS integrated circuit
Q1	2N2222 (6 V) or TIP31 (12 V) NPN power transistor (see text)
Q2	2N2907 (6 V) or TIP32 (12 V) PNP power transistor (see text)
Led1	Red common led
X1, X2, X3	Normally closed sensors (see text)
S1	SPST momentary switch
SPKR	4 Ω or 8 Ω, 4-inch loudspeaker
R1	47,000 to 100,000 Ω, 1/4 W, 5% resistor (see text)
R2	1,000 Ω (6 V) or 1,500 Ω (12 V), 1/4 W, 5% resistor
R3	1,000,000 Ω, 1/4 W, 5% resistor
C1	220 µF, 12 WVDC electrolytic capacitor
C2	0.022 µF ceramic or metal film capacitor
C3	220 µF, 16 WVDC electrolytic capacitor
C4	100 µF, 16 WVDC electrolytic capacitor

Project 90: Delayed Turn-On Alarm II (P)

Loads up to hundreds of watts can be controlled with this version of the Delayed Turn-On Alarm (Project 89). This circuit drives a relay that controls heavy-duty loads such as sirens, horns, and small appliances powered from batteries or the ac power line.

Operation is the same as described in the previous project, and a turn-on delay of about 20 seconds is obtained with the components shown in the schematic diagram. You can alter the time delay by changing both C1 and R1. C1 should be between 10 and 1,000 µF, and R1 between 10 kΩ and 1 MΩ.

The sensors are magnetic types or wires, normally closed, and several of them can be wired in series to protect large areas, as explained in Project 89.

Power comes from AA cells, a battery, or ac-to-dc converters rated to currents of 250 mA or more. As the current drain is very low when the relay is off, batteries can be used.

A schematic diagram is shown in Fig. 6.10.

Parts List: Delayed Turn-on Alarm II

IC1	4093 CMOS integrated circuit
Q1	2N2222 NPN general purpose silicon transistor
D1	1N4148 general purpose silicon diode
LED1	Red common LED
K1	6 V or 12 V relay (see text), 1 A DPDT mini relay, Radio Shack 275-249 or equivalent
S1	SPST momentary switch
R1	47,000 Ω or 100,000 Ω, 1/4 W, 5% resistor (see text)
R2	1,000 Ω (6 V) or 1,500 Ω (12 V), 1/4 W, 5% resistor (see text)
R3	1,000,000 Ω, 1/4 W, 5% resistor
R4	2,200 Ω, 1/4 W, 5% resistor
C1	220 µF, 16 WVDC electrolytic capacitor
C2	100 µF, 16 WVDC electrolytic capacitor
X1, X2, X3	Normally closed sensors (see text)

K1 depends on the power supply voltage and current drained by the load. A mini DPDT, 1 A relay (Radio Shack 275-249) can control loads

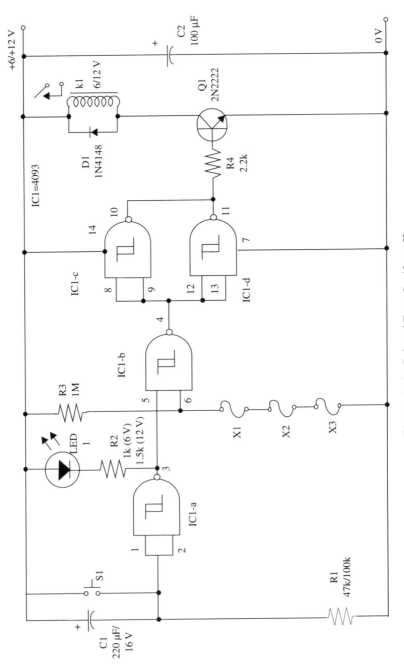

Figure 6.10 Delayed Turn-On Alarm II

up to 100 W from the ac power line. This relay uses a 12 V power supply in this project.

If you want to control heavy-duty loads, you can replace this relay with a 10 A mini SPST relay (Radio Shack 275-248), but you have to note the difference in the terminal layout.

To operate the unit, press S1 before leaving home. After the programmed time delay, the alarm will turn on automatically.

Project 91: Delayed Alarm (P) (E)

This circuit will turn on a relay after a time delay determined by C1, R4, and R3 adjustment. Time delays can range from a number of seconds to several minutes, and the sensor is a normally open switch (magnetic, microswitch, etc.). Other sensors can be wired in parallel with X1 to protect more positions. Reset is accomplished by a touch sensor or, if you prefer, an SPST momentary switch.

The circuit works as follows. When the sensor X1 is closed, IC1-a output goes low, and then its output goes high. At this moment, C1 begins to charge through R3 and R4 until IC1-c and d trigger on. After this, the transistor is biased, driving the relay (K1).

The device can be powered from 6 V or 12 V power supplies (battery or ac-to-dc converters). A 12 V, 1 A DPDT mini relay (Radio Shack 275-249) is suitable for this project.

A schematic diagram of the Delayed Alarm is shown in Fig. 6.11.

Parts List: Delayed Alarm

IC1	4093 CMOS integrated circuit
Q1	2N2907 PNP general purpose silicon transistor
D1	1N4148 general purpose silicon diode
K1	6 V or 12 V relay (see text)
X1	Magnetic or momentary switch, normally open (see text)
X2	Touch sensor or SPST momentary switch (see text)
R1	100,000 Ω, 1/4 W, 5% resistor
R2	10,000,000 Ω, 1/4 W, 5% resistor
R3	1,000,000 Ω potentiometer or trimmer pot
R4	10,000 Ω, 1/4 W, 5% resistor
R5	2,200 Ω, 1/4 W, 5% resistor
C1	10 μF to 470 μF, 16 WVDC electrolytic capacitor (see text)
C2	100 μF, 16 WVDC electrolytic capacitor

Proper positioning of the polarized components (transistor, electrolytic capacitors, and diode) must be observed.

To operate the unit, set R3 at the minimum resistance and touch X2 to reset the circuit. Then, engage X1 to trigger the alarm. After the adjusted time delay, you will hear the relay closing its contacts. Repeat this operation and, using a watch, set R3 to give the desired time delay.

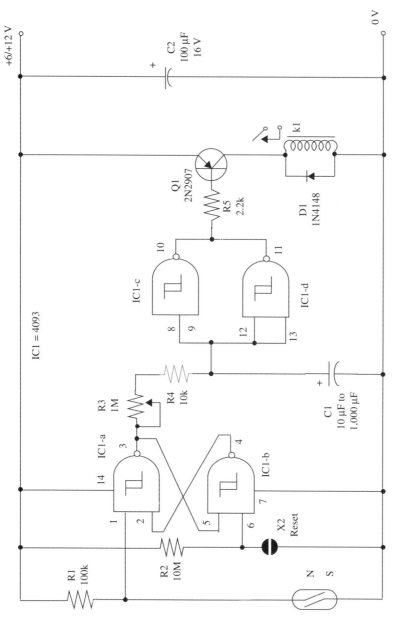

Figure 6.11 Delayed Alarm. Other sensors can be added in parallel with X1.

Project 92: Burglar Alarm Center (P)

Protect your home or business at night and over weekends with this burglar's worst enemy. When the sensors detect an intruder, the system automatically turns on an audio oscillator.

The circuit uses normally open and normally closed sensors, and you can wire dozens of them to protect your home. The device can be powered from battery or ac-to-dc converters. When the audio is off, the current drain is very low. The drain is determined by Rl and R2, and you can replace these components with 100 kΩ resistors to reduce the current drain even more. You can use either a 6 V or 12 V power supply, but you'll get more audio power with a 12 V supply.

As sensors, you can use wires, micro switches, magnetic switches (reed switches), pendulum sensors, and so on. You have to connect the sensor according its type at the appropriate input.

Figure 6.12 is schematic diagram of the Burglar Alarm Center.

Parts List: Burglar Alarm Center

IC1	4093 CMOS integrated circuit
Q1	TIP31 NPN power silicon transistor
Q2	TIP32 PNP power silicon transistor
X1 through X6	Sensors (see text)
SPKR	4 or 8 Ω, 4-inch loudspeaker
R1, R2	10,000 Ω, 1/4 W, 5% resistor
R3	47,000 Ω, 1/4 W, 5% resistor
R4	2,200 Ω, 1/4 W, 5% resistor
C1	0.022 µF ceramic or metal film capacitor
C2	220 µF, 16 WVDC electrolytic capacitor
C3	100 µF, 16 WVDC electrolytic capacitor

Transistors Q1 and Q2 should be mounted on heatsinks. The loudspeaker (SPKR) can be placed in a small enclosure to improve audio reproduction. Proper positioning of the polarized components (electrolytic capacitors and transistors) must be observed.

The parallel sensors are normally open (NO) types such as reed switches, micro switches, and others. The series sensors are wires, normally closed (NC) switches and so forth. Remember that this is a non-latching system. You can combine this project with other configurations shown in this book to get a latching alarm. Reset is accomplished by opening or closing the activated sensor.

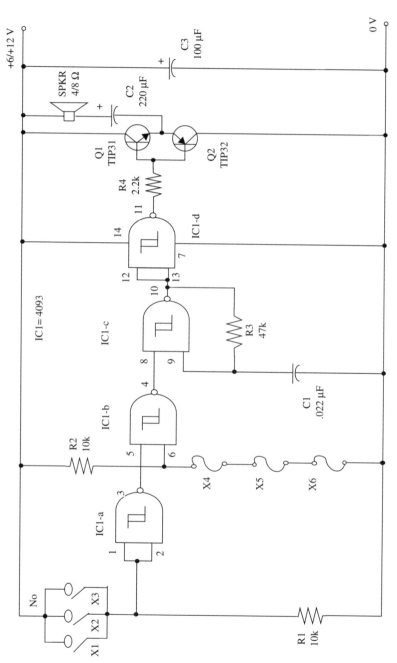

Figure 6.12 Burglar Alarm Center.

Project 93: Under-Temperature Alarm (E) (P)

This circuit will sound a piezoelectric transducer when the temperature falls below a preset point. You can use this circuit in greenhouses, heaters, and so on.

The sensor is a common silicon diode, but it suggests the possibility of another project using a thermistor. Swapping the positions of R1 and Q1 with the positions of R2 and R3, we get an over-temperature alarm.

The circuit produces an intermittent sound generated by two oscillators. IC1-a is an inverter that controls IC1-b and IC1-c. IC1-b is a very low-frequency oscillator that determines the modulation rate of the second oscillator. The second oscillator is formed by IC1-c, and its frequency is given by R5 and C2. The produced audio tone can be altered by varying both R5 and C2 within a large range of values, as described in other projects.

Power comes from a 6 V to 12 V power supply. For a portable application, you can use small 9 V batteries. The current drain is only about 0.5 mA when the tone is off. With the tone on, the current drain reaches 5 mA. Powerful output stages can be used for applications where a higher audio level is needed.

A schematic diagram of the Under-Temperature Alarm is given in Fig. 6.13.

Parts List: Under-Temperature Alarm

IC1	4093 CMOS integrated circuit
Q1	2N2222 NPN general purpose silicon transistor
D1	1N4148 general purpose silicon diode
X1	Piezoelectric transducer or crystal earpiece, Radio Shack 273-073 or equivalent
R1	10,000 Ω, 1/4 W, 5% resistor
R2	1,000,000 Ω potentiometer (trimmer)
R3	10,000 Ω,1/4 W, 5% resistor
R4	47,000 Ω, 1/4 W, 5% resistor
R5	2,200,000 Ω, 1/4 W, 5% resistor
C1	0.022 μF ceramic or metal film capacitor
C2	0.015 μF metal film or ceramic capacitor
C3	100 μF, 16 WVDC electrolytic capacitor

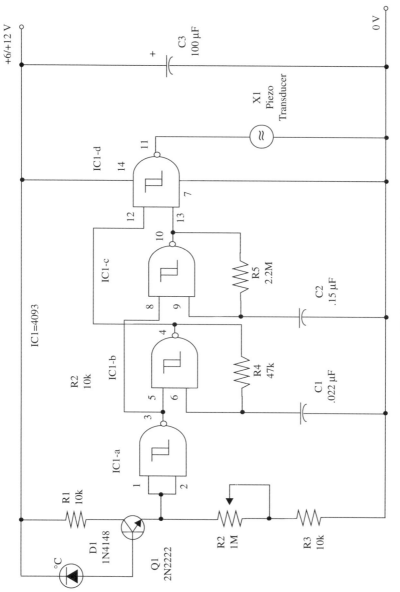

Figure 6.13   Under-Temperature Alarm.

Proper positioning of the polarized components must be observed, including diode D1, which is used as a sensor. This sensor can be placed at a distance from the device using common wires, depending on the intended application. You be careful to avoid humidity and not allow water to fall onto the sensor, which can cause problems with the circuit operation. X1 is a piezoelectric transducer or a crystal earpiece.

For more powerful audio outputs, you can use a transistorized output stage driving a loudspeaker, as described in other projects in this book.

Operation is adjusted by R2. Set this potentiometer to produce the sound at the desired temperature. For a precise adjustment, you can replace the common trimmer potentiometer with a multi-turn potentiometer.

Project 94: Under-Temperature Relay (P)

Using this project, you can turn on a heater when the temperature falls bellow a preadjusted level. You can use this device in greenhouses or in your home to keep the temperature at a desired level.

The circuit can be altered to act as an over-temperature relay simply by transposing R1, Q1, and D1 positions with R2 and R3.

To control small loads up to 1 A, a DPDT 1 A mini relay can be used (Radio Shack 275-249). But if you want to control heavy-duty loads such as powerful heaters or fans, a 10 A SPDT mini relay (Radio Shack 275-248) should be used.

Power supply selection depends on the relay used. Current drain is very low (a few milliamperes) when the relay is off. Circuit operation is the same as explained in Project 93.

A schematic diagram of the Under-Temperature Relay is shown in Fig. 6.14.

Parts List: Under-Temperature Relay

IC1	4093 CMOS integrated circuit
Q1	2N2222 NPN general purpose silicon transistor
D1, D2	1N4148 general purpose silicon diodes
K1	6 V or 12 V relay (see text)
R1, R3	10,000 Ω, 1/4 W, 5% resistor
R2	1,000,000 Ω trimmer potentiometer
R4	4,700 Ω, 1/4 W, 5% resistor
C1	100 µF, 16 WVDC electrolytic capacitor

Proper positioning of the polarized components must be observed. The sensor can be any silicon diode, and it can be placed some distance from the circuit. To adjust the unit, set R2 to close the contacts of the relay at the desired temperature.

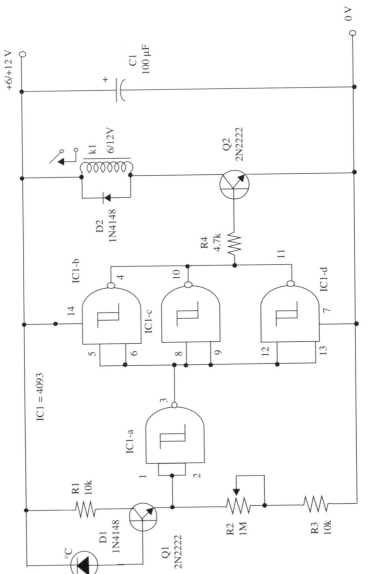

Figure 6.14 Under-Temperature Relay.

Project 95: Over-Temperature NTC Relay (P)

This circuit can be used to control the temperature of a room in the summer, turning a fan or other air circulation system on and off when the temperature reaches a preset point. Several other applications are possible, such as detection of overheating in motors, for scientific experiments, and so on. The circuit can also be used as an over-temperature alarm by connecting a siren or horn to the relay.

The circuit is powered from a 6 V or 12 V supply, depending on the relay. With 12 V supplies, you can use a 1 A mini DPDT relay such as the Radio Shack 275-249. For heavy-duty loads, the indicated part is the 10 A SPDT relay, Radio Shack 275-248.

The sensor is a negative temperature coefficient (NTC) resistor with ambient resistance between 10 and 100 kΩ. R1 and R2 values depend on the NTC resistance at ambient temperature. For NTCs between 10 and 47 kΩ, R1 is a 47 kΩ potentiometer and R1 is a 4.7 kΩ resistor. For NTCs ranging from 47 to 100 kΩ, R1 is a 100 or 220 kΩ potentiometer, and R1 is a 10 kΩ resistor.

A schematic diagram of the Over-Temperature Relay is given in Fig. 6.15.

Parts List: Over-Temperature NTC Relay

IC1	4093 CMOS integrated circuit
Q1	2N2222 NPN general purpose silicon transistor
NTC	Temperature sensor, 10 to 100 kΩ (see text)
D1	1N4148 general purpose silicon diode
K1	6 V or 12 V relay (see text)
R1	47 kΩ or 100 kΩ potentiometer (see text)
R2	4,700 Ω or 10,000 Ω, 1/4 W, 5% resistor (see text).
R3	4,700 Ω, 1/2 W, 5% resistor
C1	100 µF, 16 WVDC electrolytic capacitor

K1 is a mini DPDT relay that can be placed directly on the board. If you intend to use another type of relay, modifications in the layout should be made. The sensor is placed as far as you want from the circuit. You can place the NTC in a greenhouse and the circuit into your home, for example.

Proper positioning of the polarized components must be observed. To adjust the unit, set R1 to close the relay's contacts at the desired temperature.

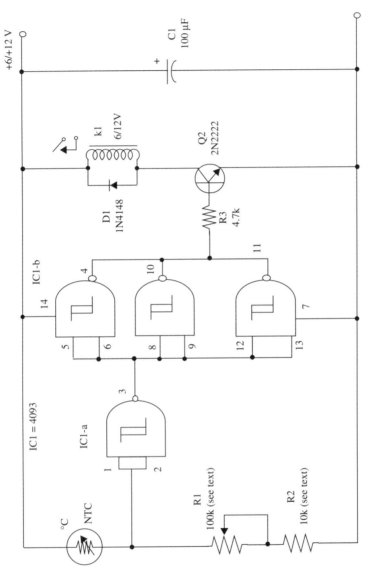

Figure 6.15 Over-Temperature NTC Relay.

Project 96: Timed Pendulum Alarm (P)

Any movement will trigger this alarm, which will operate during a preset time delay. During the time the alarm is on, an intermittent sound will be produced by a piezoelectric transducer. The on time is adjusted by R3 and can range from seconds to several minutes, depending on C3.

The modulation rate is given by R5 and C4, which control the IC1-a oscillator, and the tone frequency is determined by R6 and C5, which control the IC1-b oscillator. You can vary all these components to change the sound.

IC1-c and IC1-d act as buffers driving the piezoelectric transducer. You can replace the transducer by a transistorized output stage to drive a loudspeaker if you need more audio volume.

The circuit can be powered from 6 V to 12 V power supplies, and current drain is very low when the tone is off (only a few milliamperes). When the tone is on, the current drain is about 10 mA.

Several pendulum sensors can be wired in parallel to protect locations with many critical points. The sensors can be installed as far as you want from the control circuit.

A schematic diagram of the Timed Pendulum Alarm is shown in Fig. 6.16.

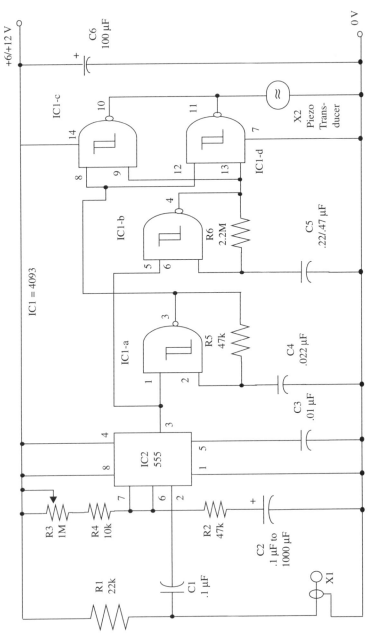

Figure 6.16 Timed Pendulum Alarm.

Parts List: Timed Pendulum-Alarm

IC1	4093 CMOS integrated circuit
IC2	555 timer integrated circuit
X1	Pendulum sensor (see text)
X2	Piezoelectric transducer or crystal earpiece, Radio Shack 273-073 or equivalent
R1	22,000 Ω, 1/4 W, 5% resistor
R2	47,000 Ω, 1/4 W, 5% resistor
R3	1,000,000 Ω potentiometer
R4	10,000 Ω, 1/4 W, 5% resistor
R5	47,000 Ω, 1/4 W, 5% resistor
R6	2,200,000 Ω, 1/4 W, 5% resistor
C1	0.1 µF ceramic or metal film capacitor
C2	1 µF to 1,000 µF, 12 WVDC electrolytic capacitor (see text)
C3	0.01 µF ceramic or metal film capacitor
C4	0.022 µF ceramic or metal film capacitor
C5	0.22 µF to 0.47 µF ceramic or metal film capacitor
C6	100 µF, 16 WVDC electrolytic capacitor

The sensor is made using a rigid bare wire passing through a bare wire ring, as shown the figure. C2 depends on the desired on period, and it can be in the range of 1 to 1,000 µF. With a 1,000 µF capacitor, the time delay is about 15 minutes (with R3 adjusted to the maximum resistance).

To adjust the device, set R3 to maximum resistance and engage the sensor. After the tone goes off, you can adjust the desired "on" duration.

7

Inverters

Inverters are circuits designed to produce high ac or dc voltages from dc power supplies such as cells, batteries, and alternative energy sources as photocells, dynamos, wind generators, and so on.

In this chapter, we will describe several inverters based on the 4093 that can be used in your home, car, scientific experiments, business applications, and in many other places.

As in the other projects, the circuits can be altered for specific applications, and we invite the reader to try all possible modifications. Better performance can also be achieved by changing several components according the variables that influence the projects' operation.

Project 97: Simple Fluorescent Lamp Inverter (E) (P)

This circuit will light a large fluorescent tube (7 to 40 W) from a 12 V power supply (e.g., car battery, NiCad cells, or other sources) without need of a choke or starter. Even old tubes that no longer function at all on the ac power line will light when used for this purpose. You can use the device as an emergency light, in trailers, for signaling, and so on.

The current drain depends on the characteristics of the lamp, and the transformer and can range from 100 to 800 mA.

The lamp brightness will depend on the drained current. Experiments should be done with several transformers and oscillator frequencies to get the best performance.

The circuit is formed by a low-frequency oscillator (IC1-a) that drives a buffer formed by IC1-b, c, and d. The buffer drives a Darlington power transistor whose load is a small transformer. High voltage is obtained from the transformer to light the fluorescent lamp.

You can alter the frequency of the oscillator by changing R1 and C1 to find the appropriate light level. If desired, you can replace R1 with a 100 kΩ potentiometer in series with a 10 kΩ resistor. This will allow frequency adjustment to obtain better performance.

A schematic diagram of the Fluorescent Lamp Inverter is shown in Fig. 7.1.

Parts List: Simple Fluorescent Lamp Inverter

IC1	4093 CMOS integrated circuit
Q1	TIP120 NPN Darlington power transistor
X1	Fluorescent lamp, 7 to 40 W (see text)
T1	117 Vac/12 Vac, 450 mA transformer, Radio Shack 273-1375 (see text)
R1	47,000 Ω, 1/4 W, 5% resistor
R2	2,200 Ω, 1/4 W, 5% resistor
C1	0.22 μF metal film or ceramic capacitor
C2	100 μF, 16 WVDC electrolytic capacitor

Transistor Q1 must be mounted on a large heatsink. T1 is a 12.6 V, 450 mA transformer. Radio Shack 273-1375 is suitable for this project but, because it has a CT transformer, you should use only two of the three secondary wires. You can experiments with a transformer with secondary currents ranging from 300 to 800 mA and voltages between 9 and 15 V for better performance.

You can also increase the power output by replacing Q1 by a power FET. Any power FET rated to 2 A or more can be used in this circuit. The IRF640 is suitable for this task.

The fluorescent lamp is rated from 7 to 40 W, and even lamps that will not function on the ac power line can be used.

Warning: the fluorescent lamp is powered with dangerous high voltage. Be careful with wire insulation and device connections.

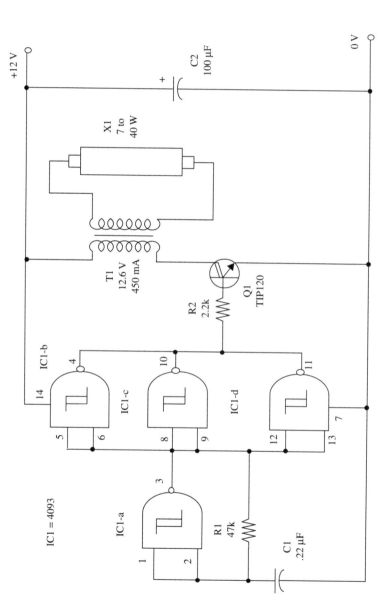

Figure 7.1   Simple Fluorescent Lamp Inverter. Any 450 mA to 1 A transformer can b used. Q1 also can use a power FET.

Project 98: Ultraviolet Lamp (P)

This battery-powered ultraviolet lamp reveals color patterns in many substances. Many substances in nature look pretty and dull in natural light but, when illuminated by an ultraviolet source, will take on the appearance of colorful gems. Minerals in rocks, sand and dirt, some insects, and common materials as cardboard and plastic fluoresce with beautiful colors when illuminated by an ultraviolet source. Interesting projects for high school science can be performed using this circuit.

This ultraviolet lantern operates with a 6 V battery formed by four D cells (alkaline or NiCad), and current drain ranges from 100 to 400 mA, depending on R1 and the characteristics of the transformer and the UV tube.

The ultraviolet source is a UV tube rated from 4 to 7 W such as the GE F6T4/BLB (6 W). All the components can be housed in a plastic box. The lamp is mounted in a PVC tube with a handle and a long cable to connect it to the circuit.

The circuit driving the lamp is a simple inverter formed by a low-frequency oscillator (IC1-a) and an output stage formed by Q1. IC1-b, c, and d act as buffers. Values of resistor R1 and capacitor C1 can be varied to achieve better performance.

A schematic diagram of the Ultraviolet Lamp is given in Fig. 7.2.

Parts List: Ultraviolet Lamp

IC1	4093 CMOS integrated circuit
Q1	TIP120 NPN Darlington power transistor
T1	12.6 V, 350 mA transformer such as Radio Shack 273-1385 (see text)
X1	Ultraviolet lamp, GE F6T4/BLB or equivalent, 4 to 7 W
S1	SPST toggle or slide switch
B1	6 V, four D cells, alkaline or NiCad
R1	39,000 Ω, 1/4 W, 5% resistor
R2	2,200 Ω, 1/4 W, 5% resistor
C1	0.22 µF ceramic or metal film capacitor
C2	100 µF, 16 WVDC electrolytic capacitor

All the components are placed on an universal printed circuit board except the power supply, S1, the transformer, and the ultraviolet tube. The printed circuit board is fixed inside a enclosure with common screws and spacers. Transistor Q1 must be mounted on a heatsink.

To use this device, close S1 to power up the circuit. You'll hear a light hum from the transformer, indicating oscillation. Some visible light will be produced by the lamp, which indicates circuit operation.

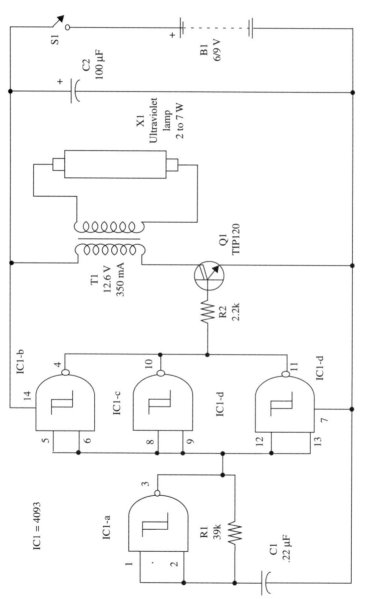

Figure 7.2 Ultraviolet Lamp.

Project 99: Experimental High-Voltage Generator (E) (P)

This circuit will generate high voltages between 2,000 and 10,000 V and can be used in several experiments and practical applications as ions generation, cattle fences, and so on.

The circuit is formed by a 4093 arranged as a low-frequency oscillator (IC1-a) and a buffer (IC1-b, c, and d) driving a Darlington power transistor (or a power FET if you prefer).

The transistor's load is a common car ignition coil that produces the desired high voltage in its secondary. R1 can be adjusted for better performance of the circuit, and the voltage depends on the coil employed.

An ac-to-dc converter can be used to power the circuit in the laboratory, and you can also use a car battery or eight D NiCad or alkaline cells. Current drain depends on R1 adjustment.

Sparks ranging from 1/10 to 1/2 inch can be obtained between HV and E terminals.

Caution: don't touch any part of the circuit when in operation.

A schematic diagram of the Experimental HV Generator is given in Fig. 7.3.

Parts List: Experimental High-Voltage Generator

IC1	4093 CMOS integrated circuit
T1	Any car ignition coil (see text)
Q1	TIP120 NPN Darlington power transistor
R1	100,000 Ω, potentiometer
R2	10,000 Ω, 1/4 W, 5% resistor
R3	2,200 Ω, 1/4 W, 5% resistor
C1	0.22 µF ceramic or metal film capacitor
C2	100 µF, 16 WVDC electrolytic capacitor

The transistor must be mounted on a heatsink. T1 is an ignition coil like the ones used in cars. Any type is suitable for this project. The positions of the polarized components must be observed.

R1 adjusts the oscillator frequency for better performance.

As an experiment, position a fluorescent lamp or neon lamp near the high voltage terminal. It will glow, and no contact is needed. A dark ambient produces better results.

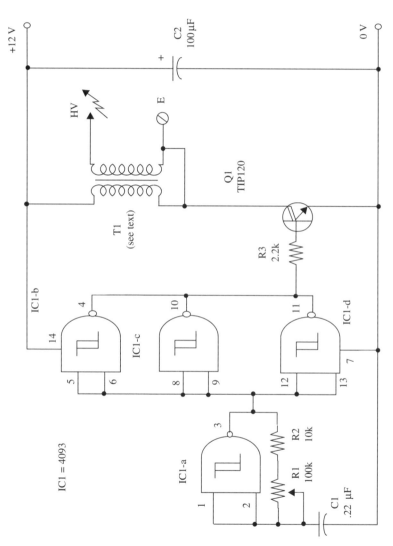

Figure 7.3 Experimental High-Voltage Generator.

Project 100: Nerve Stimulator (P)

This device can be used in controlled biological experiments. The circuit will produce high voltages (up to 300 V) at low current rates to stimulate nerves in several kind of tests.

The intensity of stimulus is adjusted by R5, and frequency is adjusted by R1. The circuit is powered from D cells (NiCad or alkaline) or a 6 V rechargeable battery. Current drain is about 100 mA. The neon lamp is used to indicate circuit operation.

The electrodes depend on the experiments. Two metal tubes with diameters ranging from 1/2 to 1 inch can be used for manual stimulation.

A schematic diagram of the Nerve Stimulator is shown in Fig. 7.4.

Parts List: Nerve Stimulator

IC1	4093 CMOS integrated circuit
Q1	TIP120 NPN Darlington power transistor
NE-1	Common neon lamp, NE-2H or equivalent
T1	12.6, 300 mA transformer, primary 117 Vac, such as Radio Shack 273-1385
R1	10,000 Ω potentiometer
R2	10,000 Ω, 1/4 W, 5% resistor
R3	2,200 Ω, 1/4 W, 5% resistor
R4	150,000 Ω, 1/4 W, 5% resistor
R5	10,000 Ω potentiometer
C1	100 µF, 16 WVDC electrolytic capacitor
C2	0.022 µF ceramic or metal film capacitor
J1	Output jack

Any small transformer with the primary rated to 117 Vac and the secondary with voltages ranging from 6 V to 12.6 V and currents between 100 mA and 500 mA can be experimented. Best results are obtained by adjusting R1. Radio Shack 273-1385 is a transformer that can be used in this project.

Transistor Q1 must be mounted on a heatsink. Any neon lamp can be used to indicate the circuit's operation. The output is a common enclosed jack (mono) and a phono plug connects the electrodes.

To use, first adjust R5 to a minimum and close S1 to power up the circuit. After this, adjust R1 and R5 for the desired stimulus.

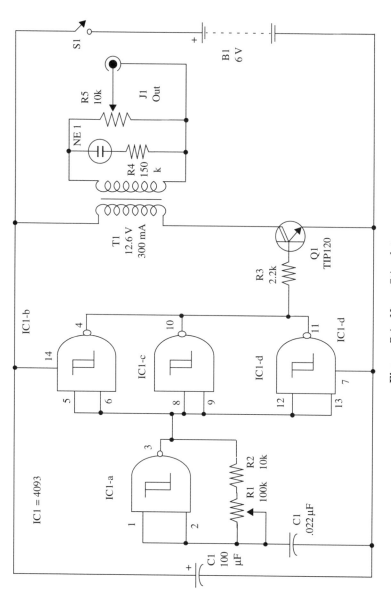

Figure 7.4 Nerve Stimulator.

Project 101: Dark-Activated Fluorescent Lamp Flasher (P)

This circuit turns on automatically at dusk and off again at dawn. It can be used in visual alarms and "attention getters" in commercial advertising.

The circuit is powered from a 12 V car battery, and current drain depends on the lamp and transformer used in the project. Fluorescent lamps ranging from 7 to 40 W can be used.

R1 adjusts the turn-on light level. R1 and the LDR are wired as a potential divider, and it is adjusted so that the voltage on IC1-a input is slight less than the gate trigger point. R3 and C2 determine the flash rate and can be altered according the intended application. C1 and R2 determine the frequency of the oscillator formed by IC1-b. Q1 drives a small transformer that generates high voltage to the fluorescent lamp.

A schematic diagram of the Dark Activated Fluorescent Flasher is given in Fig. 7.5.

Parts List: Dark-Activated Fluorescent Lamp Flasher

IC1	4093 CMOS integrated circuit
Q1	TIP120 NPN Darlington power transistor
LDR	CdS photocell, Radio Shack 276-1657 or equivalent
X1	Fluorescent lamp, 7 to 40 W (see text)
T1	12.6 V, 300 mA transformer, Radio Shack 273-1385 (see text)
F1	1 A fuse and holder
R1	1,000,000 Ω potentiometer or trimmer pot
R2	47,000 Ω, 1/4 W, 5% resistor
R3	2,200,000 Ω, 1/4 W, 5% resistor
C1	0.22 µF ceramic or metal film capacitor
C2	0.47 µF ceramic or metal film capacitor
C3	100 µF, 16 WVDC electrolytic capacitor

Transistor Q1 must be mounted on a heatsink. The board, transformer, F1, and R1 are housed in a plastic box. The LDR (sensor) should be mounted in a place that receives the ambient light but not the lamp flashes.

The fluorescent lamp can be installed as far as you want from the device, but you must take care with the wiring insulation, because high voltages are present in this part of the circuit. Change the values of C1 and C2 to get better performance and a flash rate that is appropriate for the application you have in mind.

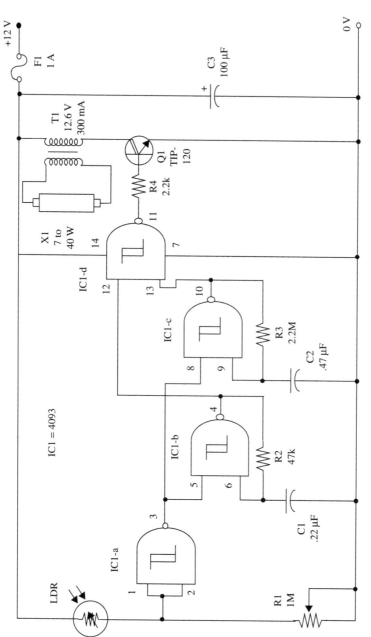

Figure 7.5 Dark-Activated Fluorescent Lamp Flasher.

Project 102: Light-Activated Fluorescent Lamp Flasher (P)

This circuit will turn on a fluorescent light when the sensor (LDR) is illuminated. The circuit can be used as a visual alarm and in other applications for home and business.

Operation is the same as Project 101 except as affected by the transposition of the LDR with the potentiometer R1.

The circuit is powered from a 12 V car battery, and the fluorescent lamp can range from 7 to 40 W. The flash rate can be altered by changing C2 and R3.

Current drain depends on the transformer and the fluorescent lamp, ranging from 100 mA to 400 mA.

The sensor and the fluorescent lamp can be installed as far as you want from the device, but they should be separated to avoid feedback.

A schematic diagram of the Light Activated Fluorescent Flasher is shown in Fig. 7.6.

Parts List: Light-Activated Fluorescent Lamp Flasher

IC1	4093 CMOS integrated circuit
Q1	TIP120 NPN Darlington power transistor
LDR	CdS photocell, Radio Shack 276-1657 or equivalent
T1	12.6 V, 300 mA, Radio Shack 273-1385 or equivalent transformer, primary 117 Vac
X1	7 to 40 W fluorescent tube
R1	1,000,000 Ω potentiometer or trimmer pot
R2	47,000 Ω, 1/4 W, 5% resistor
R3	2,200,000 Ω, 1/4 W, 5% resistor
R4	2,200 Ω, 1/4 W, 5% resistor
C1	0.22 μF ceramic or metal film capacitor
C2	0.47 μF ceramic or metal film capacitor
C3	100 μF, 16 WVDC electrolytic capacitor

The positions of the polarized components (electrolytic capacitor, integrated circuit, and transistor) must be observed. Transistor Q1 must be mounted on a heatsink.

The power supply is a rechargeable battery (NiCad, for instance) or a car battery. The transformer is the same as the one used in Project 101. Installation details are as given in that project.

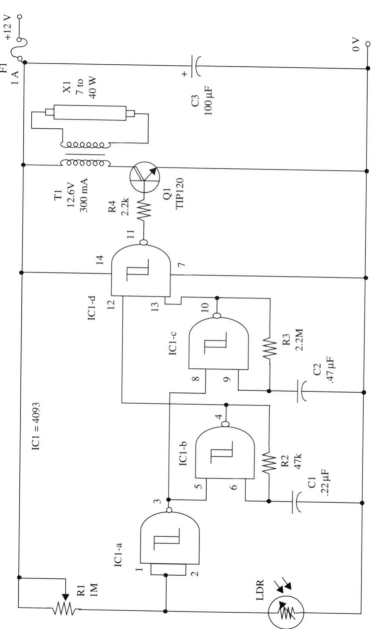

Figure 7.6   Light-Activated Fluorescent Lamp Flasher.

Project 103: Fluorescent Lamp Flasher (P) (E)

The inverter shown in this project will flash a fluorescent lamp from a 12 Vdc power supply such as a car battery or NiCad battery. The circuit can be used alone in visual advertising or decoration or with other circuits for home or business applications.

The flash rate depends on R2 and C2, which can be altered to change the circuit performance. Current drain, and therefore the lamp brightness, depends on the characteristics of the transformer and lamp. Current drain is between 100 and 400 mA.

Any small transformer rated for 5 to 12 V and currents between 100 and 500 mA can be used in this project. Radio Shack 273-1385 is suitable.

A schematic diagram of the Fluorescent Lamp Flasher is given in Fig. 7.7.

Parts List: Fluorescent Lamp Flasher

IC1	4093 CMOS integrated circuit
Q1	TIP120 or equivalent NPN Darlington power transistor
F1	1 A fuse and holder
X1	7 to 40 W fluorescent lamp
T1	12.6 V, 300 mA secondary, 117 Vac primary transformer (see text)
R1	47,000 Ω, 1/4 W, 5% resistor
R2	2,200,000 Ω, 1/4 W, 5% resistor
R3	2,200 Ω, 1/4 W, 5% resistor
C1	0.22 µF ceramic or metal film capacitor
C2	0.47 µF ceramic or metal film capacitor
C3	100 µF, 16 WVDC electrolytic capacitor

The transformer is the same as used in Projects 102 and 103 (see these projects for more details). Transistor Q1 must be mounted on a heatsink.

The fluorescent lamp is rated from 7 to 40 W, and even ones that no longer function at power line voltage can be used.

Better performance is achieved by adjusting the values of capacitor C1 and resistor R1. These components can be altered in a wide range of values. You can also replace Q1 by any power FET to obtain better performance.

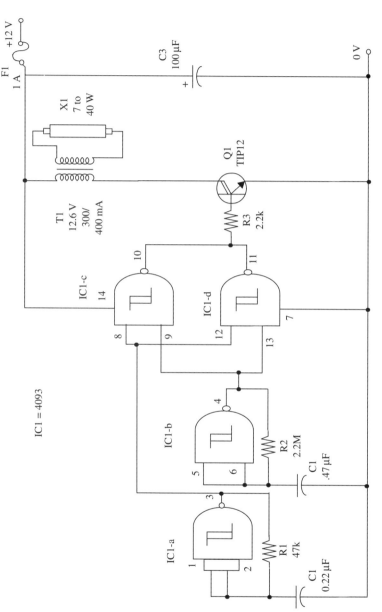

Figure 7.7 Fluorescent Lamp Flasher

Project 104: Negative Ion Generator (P)

Negative and positive ions in the ambient affect human behavior. Many theories explain what happens, and some attribute irritability and erratic behavior to positive ions, and feelings of well being to negative ions. Numerous products exist in the market to flood a home or business with negative ions.

The circuit we present here generates negative ions and can be used in experiments to determine their effects. As the ion level varies according the used components, we do not recommend that the reader use this project at a home or business until making a precise measurement of the amount of ions generated by it.

The circuit is basically a high-voltage generator that produces, via an electrode, a constant flux of negative ions. These ions are dispersed into the ambient air, affecting living beings in the vicinity.

In the circuit, IC1-a acts as a low-frequency oscillator that drives a buffer (IC1-b, c, and d) and a power output transistor. The transistor has as its load a small transformer that generates a high ac voltage of about 150 V. This high voltage is applied to a voltage multiplier that produces an output about of 2 kV. This voltage is enough to produce a constant flux of ions through the electrode X1.

The circuit is powered from a 12 V source as an ac-to-dc converter or a battery. Current drain depends on the components and typically is between 100 and 500 mA.

A schematic diagram of the Negative Ion Generator is shown in Fig. 7.8.

Parts List: Negative Ion Generator

IC1	4093 CMOS integrated circuit
Q1	TIP120 NPN Darlington power transistor
D1–D13	1N4007 (1 A, 800 V) silicon rectifier diodes
T1	12.6 V, 300 mA transformer, Radio Shack 273-1385 or equivalent, primary 117 Vac
X1	Electrode (see text)
R1	39,000 Ω, 1/4 W, 5% resistor
R2	4,700 Ω, 1/4 W, 5% resistor
R3, R4, R5	2,200,000 Ω, 1/4 W, 5% resistors
C1	0.22 µF ceramic or metal film capacitor
C2–C14	0.01 µF, 630 WVDC ceramic or metal film capacitors
C15	100 µF, 16 WVDC electrolytic capacitor

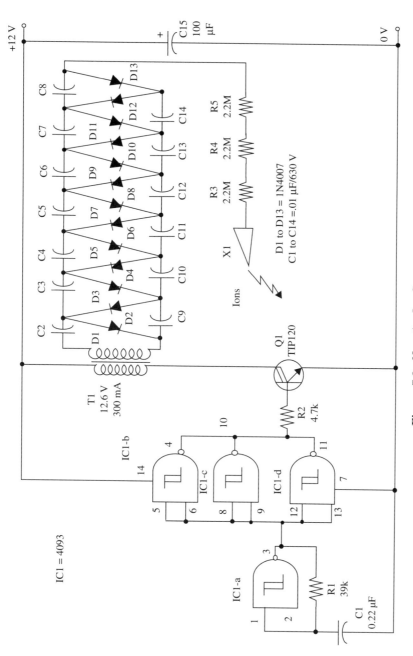

Figure 7.8 Negative Ion Generator.

Electrode X1 is a small pin. Any small transformer with a secondary rated for 5 to 12 V and currents between 100 and 500 mA can be used experimentally. The primary wound is for 117 Vac. Transistor Q1 must be mounted on a small heatsink.

Project 105: Fluorescent Strobe Light (P) (E)

Rapid light pulses of short duration produce interesting effects when used to illuminate objects in continuous movement. The movements are "frozen," as we can observe in musical shows, dances, and so on.

This circuit drives a common florescent lamp (7 to 40 W), which is different from the xenon tubes e normally used in this kind of project. Fluorescent tubes are easy to locate, but they do not produce light pulses with the same intensity as a xenon tube. Therefore, our project should be considered a "low-power experimental strobe light," but you can use it in your home, for decoration, and in many other applications.

The circuit consists of a simple modulated inverter. IC1-a acts as a low-frequency oscillator driving a transistorized output stage via a buffer. The transistor has as its load a small transformer that drives a fluorescent lamp with high-voltage, short-duration pulses.

The circuit can be powered from any 12 V power supply. Current drain is between 100 and 500 mA, depending on the components used in the project.

A schematic diagram of the Fluorescent Strobe Light is shown in Fig. 7.9.

Parts List: Strobe Fluorescent Light

IC1	4093 CMOS integrated circuit
Q1	TIP120 NPN Darlington power transistor
X1	7 to 40 W common fluorescent lamp
T1	12.6 V, 300 mA or any small transformer (see text) such as Radio Shack 273-1385, primary 117 Vac
R1	47,000 Ω, 1/4 W, 5% resistor
R2	4,700,000 Ω potentiometer
R3	1,000,000 Ω, 1/4 W, 5% resistor
R4	2,200 Ω, 1/4 W, 5% resistor
C1	0.22 µF ceramic or metal film capacitor
C2	0.47 µF ceramic or metal film capacitor
C3	100 µF, 16 WVDC electrolytic capacitor

Transistor Q1 must be mounted on a small heatsink. Any small transformer with secondary voltages between 5 and 12 V can be used in this project.

Light pulses are adjusted by R2. The fluorescent lamp is wired as far as you want from the device, but *you have to take care with the wire insulation.* This part of the circuit is submitted to high voltages, and this represents *a possibility of serious shock.*

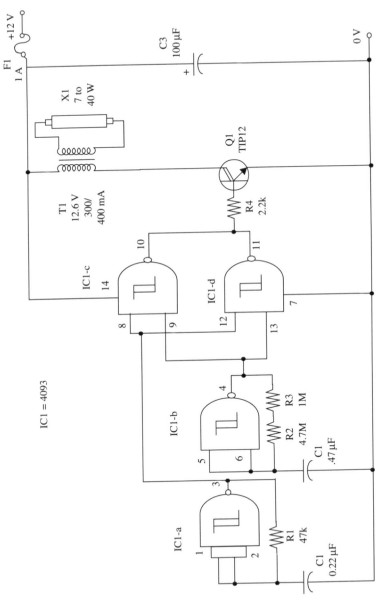

Figure 7.9 Fluorescent Strobe Light.

8

Miscellaneous Projects

The projects in this final chapter are intended for many applications for the home, school science projects, scientific labs, amateur science labs, cars, and businesses. Many of them can be used as part of more complex projects. Light effects, sound generators, timers, testers, and musical instruments are included in this chapter. Many projects will include other integrated circuits, but all are based on the 4093.

Special attention should be given by the reader to the projects powered from the ac power line. Take care with all connections, and verify the insulation to avoid shocks and shorts.

As in previous chapters, these projects can be modified to obtain better performance, depending on the components you use and the intended application. Many experiments can be made with several components, as indicated in each case.

Project 106: Gated Oscillator I (E)

This experimental project will teach you how we can use a 4093 as a gated oscillator. The oscillator is gated by an external pulse. You can mount it to experiment with the configuration or to use it as part of a more complex project.

The circuit is gated by a positive-moving pulse applied to the input. The frequency is given by resistance (R) and capacitance (C), according to the description given in Chapter 1. Capacitance can range from 50 pF to 1,000 µF, and resistance can range between 2 kΩ and 4.7 MΩ.

The circuit is powered from 5 to 15 V supplies, and current drain without load is approx. 0.5 mA.

A schematic diagram of Gated Oscillator I is given in Fig. 8.1.

The gate pulse voltage shouldn't exceed the power supply voltage. Low level is 0 V. Input impedance is very high, approximately several megohms, and the output characteristics are as given in Chapter 1.

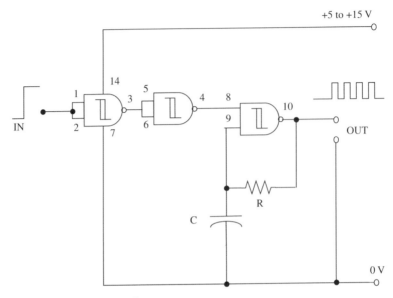

Figure 8.1   Gated Oscillator I.

Parts List: Gated Oscillator I

IC1	4093 CMOS integrated circuit
R	2,000 Ω to 4,700,000 Ω resistor (see text)
C	50 pF to 1,000 µF capacitor (see text)

Project 107: Gated Oscillator II (E)

This circuit triggers on with a negative-moving pulse at its input. We use only two of the four 4093 IC gates, and other characteristics are the same as in Project 106.

You can use this circuit as part of more complex ones. A schematic diagram of Gated Oscillator II is given in Fig. 8.2.

Parts List: Gated Oscillator II

IC1	4093 CMOS integrated circuit
C	50 pF to 1,000 µF capacitor (see Project 106)
R	2,000 to 4,700,000 Ω resistor (see Project 106)

The frequency is given by R and C. See Chapter 1 to determine these components' values.

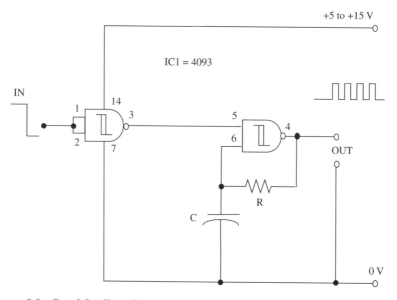

Figure 8.2   Gated Oscillator II.

Project 108: 60 Hz Generator (E)

This circuit produces a precise 60 Hz squarewave output that can be used to drive clocks, computers, timers, an many other projects. The frequency is given by the ac power line, and very few components are needed to complete the project.

A schematic diagram of the 60 Hz generator is shown in Fig. 8.3.

Parts List: 60 Hz Generator

IC1	4093 CMOS integrated circuit
T1	6,3 V, 300 mA transformer (see text)
R1	100,000 Ω trimmer pot
C1	1,000 μF, 25 WVDC electrolytic capacitor
D1, D2	1N4002 or equivalent silicon rectifier diodes

Any transformer rated for 5 to 12 V and currents ranging from 50 mA to 1 A (or as required by the project) can be used. The circuit can also be used to power any project driven by a 60 Hz generator—a clock, for instance. R1 is adjusted to give the required output. An oscilloscope can be used to adjust the circuit.

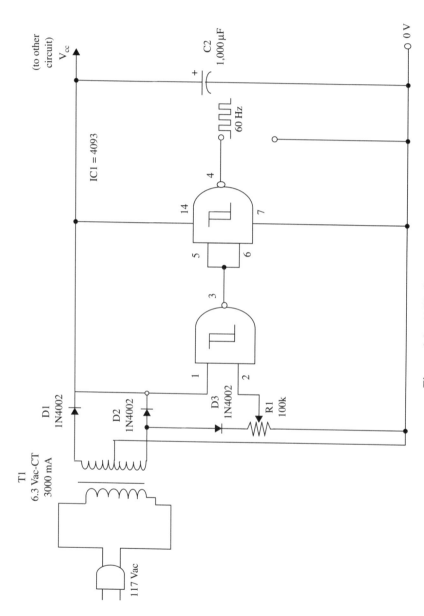

Figure 8.3 60 Hz Generator.

Project 109: Electroscope (P)

This circuit can be used to detect static electricity and also for "sniffing" high voltage without actually making contact with dangerous circuitry. A metal ring, which forms the sensor, is simply poked into the electric field. *This operation must be done with caution to avoid contact with conductors carrying the high voltage.*

You can use the device in experiments with high-voltage generators and many other applications at home and in school scientific projects and experiments. The circuit is portable and it is powered from four AA cells or a 9 V battery.

The presence of static charges or electric high-voltage fields is indicated by an LED.

A schematic diagram of the electroscope is shown in Fig. 8.4.

Parts List: Electroscope

IC1	4093 CMOS integrated circuit
LED1	Red common LED
S1	SPST slide or toggle switch
B1	6 V or 9 V (four AA cells or 9 V battery)
X1	Sensor (see text)
C1	1 pF ceramic capacitor
C2	100 µF, 12 WVDC electrolytic capacitor
R1	1,000 Ω, 1/4 W, 5% resistor

The sensor is a small wire ring. The circuit can be housed into a small plastic box. Capacitor C1 can be a "home-made" unit. You just have to twist two 2-inch solid wire pieces to form a capacitor.

To operate, bring a charged source close to the sensor. A piece of paper stroked with a rod if insulating material will do. The LED will glow according the charge movement.

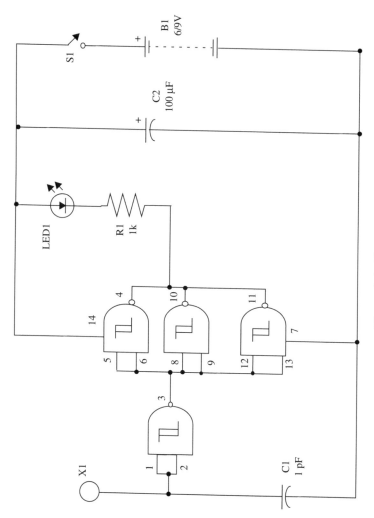

Figure 8.4   Electroscope.

Project 110: Modulated Generator (E)

This circuit is intended for laboratory applications. It produces a continuous modulated square wave ranging from 100 Hz to 1 kHz (or in another range, with component modifications).

The output is adjusted from 0 to 6 or 9 V by R5, and the frequency is adjusted by R1. Modulation is adjusted by R3.

The circuit works as follows. IC1-a is wired as an audio oscillator, and the frequency is given by C1, R1, and R2. You can vary the value of C1 between 1 and 100 nF to modify the frequency range.

IC1-b is a very low-frequency oscillator that acts as a modulator. Frequency is adjusted by R3. S2 is used to select the output signal. When S2 is in (a) we have a modulated signal at the output. When S2 is in (b) we have a continuous signal at the output. R5 adjusts the output signal's amplitude.

A schematic diagram of the Modulated Generator is shown in Fig. 8.5.

Parts List: Modulated Generator

IC1	4093 CMOS integrated circuit
S1	SPST toggle or slide switch
S2	SPDT toggle or slide switch
B1	6 or 9 V, four AA cells or battery
R1	100,000 Ω potentiometer
R2	10,000 Ω, 1/4 W, 5% resistor
R3	2,200,000 Ω potentiometer
R4	100,000 Ω, 1/4 W, 5% resistor
R5	10,000 Ω potentiometer
C1	0.022 μF ceramic or metal film capacitor
C2	0.22 or 0.47 μF ceramic or metal film capacitor
C3	0.01 μF ceramic or metal film capacitor
C4	100 μF, 12 WVDC electrolytic capacitor
J1	Mono enclosed jack

The circuit can be housed into a small plastic box. Frequency calibration can be made with a frequency meter or oscilloscope.

C1 and C2 can be varied to change the frequency and modulation range. The circuit can be powered from 6 or 9 V power supplies, and current drain is only few milliamperes.

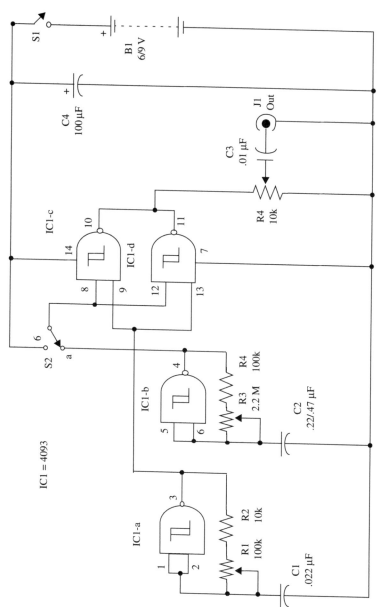

Figure 8.5 Modulated Generator.

Project 111: Capacitor Tester (P)

Capacitors ranging from 120 pF to 100 µF (any type) can be tested with this simple circuit. Indication of state is visual. By adjusting R1, you will find a point at which the LED flashes. If the LED stays on, the capacitor is shorted, and if the LED stays off, the capacitor is open.

The circuit can be powered from 6 V (four AA cells) or 9 V (battery) supplies, and current drain is only about 0.5 mA with the LED off. With the LED on, current drain is about 5 mA.

The circuit consists of a very low-frequency oscillator (IC1-a) adjusted by R1. This potentiometer should be adjusted so that the frequency goes to approximately 1 Hz when the LED's flash can be observed. IC1-b, c, and d act as simple buffers to drive the LED.

A schematic diagram of the Capacitor Tester is given in Fig. 8.6.

Parts List: Capacitor Tester

IC1	4093 CMOS integrated circuit
LED1	Red common LED
S1	SPST toggle or slide switch
B1	6 V or 9 V, four AA cells or battery
X1, X2	Red and black probes or alligator clips
R1	2,200,000 Ω potentiometer
R2	10,000 Ω, 1/4 W, 5% resistor
R3	1,000 Ω, 1/4 W, 5% resistor
C1	100 µF, 12 WVDC electrolytic capacitor

The project can be housed into a small plastic box. Positions of the polarized components (LED, electrolytic capacitor, etc.) should be observed.

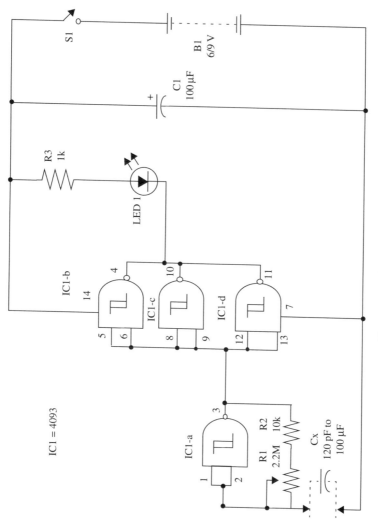

Figure 8.6 Capacitor Tester.

Project 112: X-Ray Detector (P) (E)

X-ray exposure is dangerous to human beings. This circuit can detect sources of X-rays, producing an audible sound in their presence. The circuit can be used in the laboratory, home, and other places where X-rays can be present.

The circuit works as follows: the sensor is an LDR (CdS photocell) whose resistance depends on the amount of light that falls onto a sensitive surface. If the LDR is housed into an aluminum foil box with a small piece of fluorescent material, light doesn't pass through it, but the X-rays have no problem reaching the fluorescent material. The fluorescent material absorbs the X-rays and converts their energy in visible light. The fluorescent material glows with a weak light that excites the LDR reducing its electrical resistance.

In complete darkness, LDR resistance is very high (in the range of millions of ohms), and then the oscillator formed by IC1-a operates in a very low frequency. Only interval pulses can be produced.

In the presence of an X-ray source, the LDR resistance falls, and the oscillator increases its frequency, indicating that the X-rays are present.

A schematic diagram of the X-ray Detector is given in Fig. 8.7.

Parts List: X-Ray Detector

IC1	4093 CMOS integrated circuit
X1	Piezoelectric transducer or crystal earpiece, Radio Shack 273-073 or equivalent
LDR	CdS Photocell, Radio Shack 276-1657 or equivalent
C1	1,200 pF ceramic capacitor
C2	100 µF, 12 WVDC electrolytic capacitor

The sensor's details are given in the same figure. This sensor is connected to the circuit with common wires 4 to 20 inches in length.

The power supply can range from 6 to 12 V. If portable use is desired, you can use a 9 V battery or four AA cells.

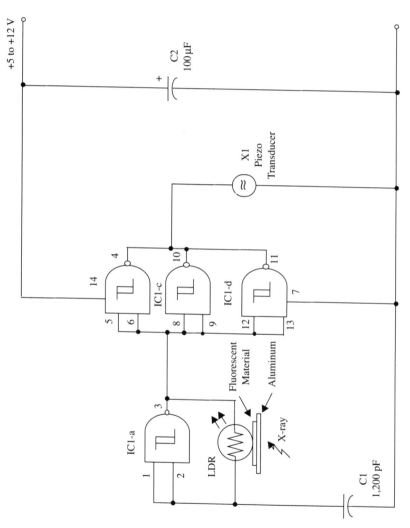

Figure 8.7 X-Ray Detector. See text for sensor details.

Project 113: Continuity Tester (P)

This circuit can be used to test electronic components as diodes, resistors, coils, transformers, lamps, fuses, switches, loudspeakers, transistors, and many other parts, producing a visual indication of state.

Two LEDs are used to indicate the state of the probed component. When the probes are separated, or when the resistance between them is very high, LED1 is on, and LED2 is off. With a low resistance between the probes (the value to be considered as low or high is fixed by the R2 adjustment), LED2 is on, and LED1 is off. The circuit can be powered from 6 to 9 V supplies such as a battery or four AA cells.

Current drain is about 10 mA, depending on R3 and R4. These resistors have varying values according the power supply voltage.

A schematic diagram of the Continuity Tester is given in Fig. 8.8.

Parts List: Continuity Tester

IC1	4093 CMOS integrated circuit
LED1, LED2	Red common LEDs
P1, P2	Probes
S1	SPST toggle or slide switch
B1	6 or 9 V battery or AA cells
R1	10,000 Ω, 1/4 W, 5% resistor
R2	1,000,000 Ω potentiometer
R3, R4	470 Ω (6 V) or 680 Ω (9 V), 1/4 W, 5% resistors
C1	100 µF, 12 WVDC electrolytic capacitor

Observe proper positioning of the polarized components (e.g., LEDs and electrolytic capacitor). To use this tester, use R2 to adjust the sensitivity. With R2 in a high-value position, the turn-on point of LED 2 will occur with high resistances. Adjust R2 according the transition resistance you want to detect.

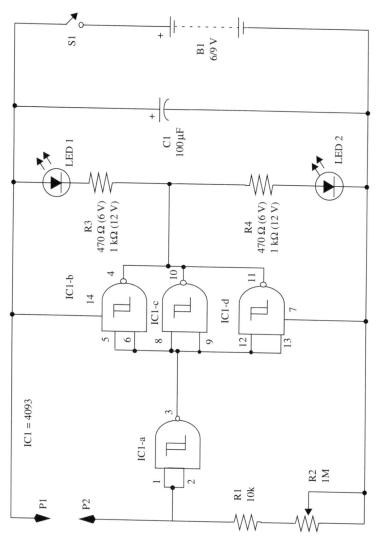

Figure 8.8 Continuity Tester.

Project 114: Auditory Capacitor Tester (P)

Capacitors in the range between 1 nF and 100 µF can be tested with this simple circuit. A piezoelectric transducer produces pulses or continuous tones when a good capacitor is probed. If the capacitor is bad, no sound will be produced.

The project is very simple, and all components can be housed in a small plastic box. The circuit is powered from 6 or 9 V power supplies, and current drain is very low. When the sound is off, current drain is about 0.5 mA, and when the sound is on, current drain is about 5 mA.

A schematic diagram of the Auditory Capacitor Tester is shown in Fig. 8.9.

Parts List: Auditory Capacitor Tester

IC1	4093 CMOS integrated circuit
S1	SPST toggle or slide switch
B1	6 or 9 V, four AA cells or battery
P1, P2	Black and red probes
X1	Piezoelectric transducer or crystal earpiece, Radio Shack 273-073 or equivalent
R1	100,000 Ω potentiometer
R2	10,000 Ω, 1/4 W, 5% resistor
C1	100 µF, 12 WVDC electrolytic capacitor

To use, touch the capacitor leads with the test probes and adjust R1 to get a continuous sound or pulses (continuous sounds are produced by small capacitors, and pulsed tones are produced by large capacitors). If no sound is produced in any position of R1, the probed capacitor is bad. (No sound is produced if the capacitor is out of range.)

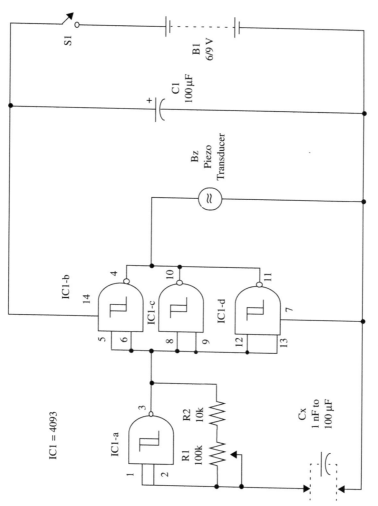

Figure 8.9 Auditory Capacitor Tester

Project 115: Fixed-Tone Continuity Tester (P)

You can probe small components as in Project 113 with this simple circuit. Components such as diodes, lamps, fuses, coils, switches, and many others can be tested with this tester, which produces a continuous tone with low-resistance devices.

The circuit can be powered from 6 or 9 V supplies, and current drain is very low. With the tone off, current drain is only 0.5 mA, and when the tone is on, current drain is about 5 mA. R1 adjusts sensitivity, as described in Project 113.

A schematic diagram of the Fixed-Tone Continuity Tester is shown in Fig. 8.10.

Parts List: Fixed-Tone Continuity Tester

IC1	4093 CMOS integrated circuit
X1	Piezoelectric transducer or crystal earpiece, Radio Shack 273-073 or equivalent
S1	SPST toggle or slide switch
B1	6 or 9 V, four AA cells or battery
P1, P2	Black and red probes
R1	1,000,000 Ω potentiometer
R2	10,000 Ω, 1/4 W, 5% resistor
R3	39,000 Ω, 1/4 W, 5% resistor
C1	0.022 µF ceramic or metal film capacitor
C2	100 µF, 12 WVDC electrolytic capacitor

All of the components can be housed into a small plastic box. R1 adjusts sensitivity, as in Project 113 (see Project 113 for more details). The circuit can be powered from AA cells or a battery. X1 is a piezoelectric transducer or crystal earpiece.

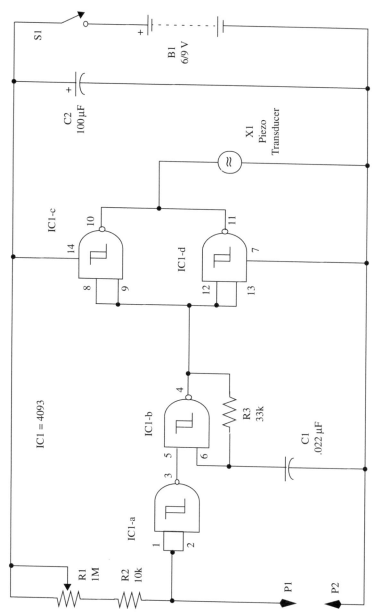

Figure 8.10 Fixed-Tone Continuity Tester.

Project 116: Time-Delayed Generator I (E) (P)

This circuit produces pulses during a programmable time delay. You can use it as part of other projects such as games (to produce a random number of pulses), for audio effects, and so on.

The time delay depends on R1 and C1 and can range from 0.1 seconds to more than 15 minutes. C1 and C2 values are given in the schematic diagram.

R2 and C2 determine the frequency of the produced pulses, and these components can be varied within a wide range of values. With values given in the schematic diagram, the frequency is about 1 Hz. You can vary these components to produce pulses in the range between 0.01 Hz and 1 MHz.

A schematic diagram of the Time-Delayed Generator is given in Fig. 8.11.

Parts List: Time-Delayed Generator I

IC1	4093 CMOS integrated circuit
S1	SPST momentary switch
R1	100,000 to 1,000,000 Ω, 1/4 W, 5% resistor (see text)
R2	10,000 to 2,200,000 Ω, 1/4 W, 5% resistor (see text)
C1	1 µF to 1,000 µF capacitor (see text)
C2	1,200 pF to 1 µF metal film or ceramic capacitor (see text)
C3	100 µF, 16 WVDC electrolytic capacitor

The circuit is powered from voltages source between 3 and 15 V, and current drain is between 0.5 and 1 mA (unloaded output).

To operate, press S1 to generate an output signal that occurs during a time period determined by the values of components employed.

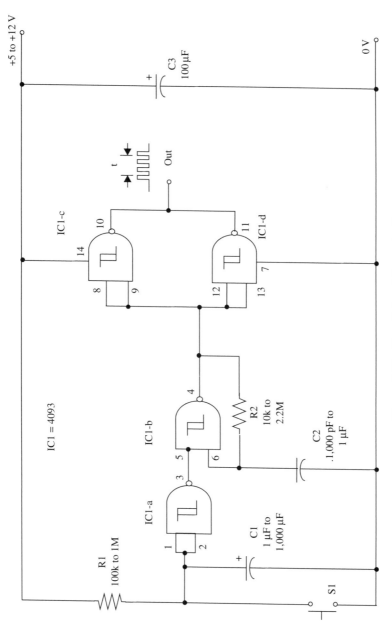

Figure 8.11   Time-Delayed Generator I

Project 117: Time-Delayed Generator II (P)

This circuit can be used as part of other projects, as explained in Project 117, or alone as a probe instrument for digital equipment. The circuit generates a square wave whose frequency is determined by R3 and C3 during a time period given by R2 and C1.

The time period is fixed between 0.1 seconds and 15 minutes, and output signal has frequencies between 0.1Hz and 1 MHz.

The circuit can be powered from voltages sources ranging from 5 to 15 V, and current drain is very low.

A schematic diagram of the Time-Delayed Generator II is shown in Fig. 8.12.

Parts List: Time-Delayed Generator II

IC1	555 timer integrated circuit
IC2	4093 CMOS integrated circuit
R1	22,000 Ω to 47,000 Ω, 1/4 W, 5% resistor
R2	10,000 Ω to 1,000,000 Ω, 1/4 W, 5% resistor
R3	10,000 to 2,200,000 Ω, 1/4 W, 5% resistor
C1	0.1 to 1,000 µF capacitor (see text)
C2	0.01 µF ceramic or metal film capacitor
C3	0.01 µF to 1 µF capacitor (see text)
C4	100 µF, 16 WVDC electrolytic capacitor

The circuit works as follows. When a negative-traveling pulse is applied to the input of IC1 (pin 2), the output goes high. This triggers "on" the oscillator formed by IC2-c. The generated square signal is applied to a buffer (IC2-d) and then to the output. IC2-a and IC2-b also act as buffers.

The "on" time depends on R2 and C1 and can range between the values shown in the schematic diagram. Times between 0.1 seconds and 15 minutes can be obtained. Output frequency is given by R3 and C3, and the values are shown in the schematic diagram.

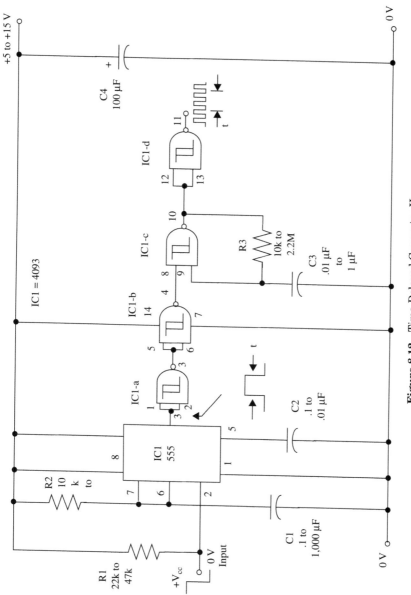

Figure 8.12   Time-Delayed Generator II

Project 118: Dexterity Tester (P)

This circuit puts your manual dexterity to a lively test. As shown in the schematic diagram, the circuit has a small loop that you need to pass around a wire. The object of the game is to guide the loop over the weaving course without touching the wire.

A slight misjudgment or quiver of the hand, and the ring will contact the weaving wire. This turns on the circuit, which will produce an audible tone. Audio tone duration is constant and is not dependent on contact between the wires.

The skill required to play the game depends largely on the size of the loop and the degree of twist and turn in the wire.

Scoring is a matter of counting the number of times the sound is produced by the device. The person with the lowest total wins.

The circuit is powered from a 9 V battery or four AA cells. A schematic diagram is shown in Fig. 8.13.

Parts List: Dexterity Tester

IC1	4093 CMOS integrated circuit
R1	1,000,000 Ω, 1/4 W, 5%
R2	47,000 Ω, 1/4 W, 5%
X1	Weaving wire (see text)
X2	Loop (see text)
X3	Piezoelectric transducer or crystal earpiece, Radio Shack 273-073 or equivalent
S1	SPST toggle or slide switch
B1	6 or 9 V, AA cells or battery
C1	1 µF to 10 µF, 12 WVDC electrolytic capacitor (see text)
C2	0.022 µF ceramic or metal film capacitor
C3	100 µF, 12 WVDC electrolytic capacitor

A printed circuit board can used to mount the components and housed in a small plastic box with the battery and transducer. The wires running to the loop and weaving wire can be 2 to 3 feet long.

Time on depends on C1, which can be varied among the values in the parts list. The circuit is powered from a 9 V battery or four AA cells. Current drain is about 0.5mA with the tone off and about 5 mA with the tone on.

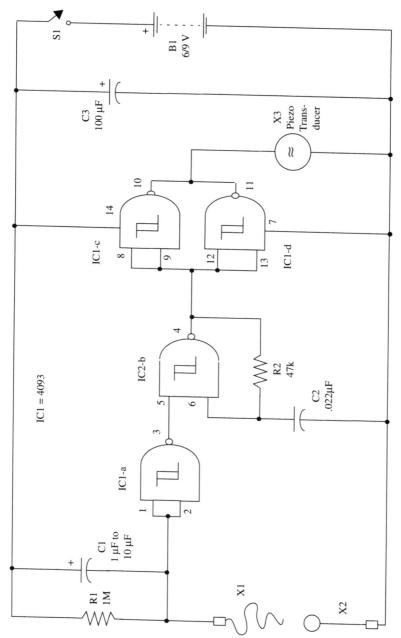

Figure 8.13 Dexterity Tester.

Project 119: Water Sensor (P) (E)

This circuit produces a continuous sound when the water level rises high enough to touch the sensor. The circuit can be used at home, in the laboratory, or business to indicate when water reaches a predetermined level in a reservoir, or to indicate rain.

The device can be powered from 6 to 12 V power supplies, and current drain is very low—only 0.5 mA when the sound is off.

A schematic diagram of the water sensor is shown in Fig. 8.14.

Parts List: Water Sensor

IC1	4093 CMOS integrated circuit
X1, X2	Water sensor (see text)
X3	Piezoelectric transducer or crystal earpiece, Radio Shack 273-073 or equivalent
R1	2,200,000 Ω, 1/4 W, 5% resistor
R2	39,000 Ω, 1/4 W, 5% resistor
C1	0.022 µF ceramic or metal film capacitor
C2	100 µF, 16 WVDC electrolytic capacitor

The sensor is made with two bare wires that touch the water, thereby triggering the circuit on. This sensor should be positioned at the level at which the reader wants the alarm to sound.

X3 is a piezoelectric transducer or a crystal earpiece, but if you want a higher audio level, use a transistorized output stage to drive a loudspeaker, as shown in other projects in this book.

The frequency of the audio tone is determined by R2 and C1. You can alter this frequency by changing the value of R2. This resistor can assume values between 22 and 100 kΩ. For an adjustable tone, replace R2 with a 100 kΩ potentiometer and a 10 kΩ series resistor.

Wires to the sensor can be long. The power supply is formed by four AA cells, a 9 V battery, or an ac-to-dc converter. *Don't use a transformerless supply in this project!*

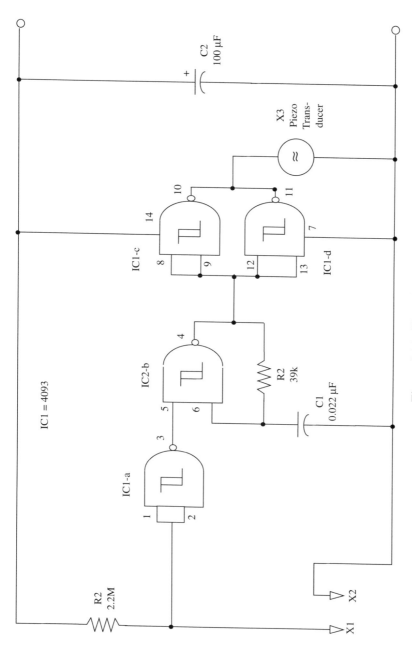

Figure 8.14 Water Sensor.

Project 120: Electronic Organ (E) (P)

This is an interesting experiment in audio that would make a nice toy for the children. This circuit produces a musical note when you touch a keyboard with a probe. The sound is produced by a piezoelectric transducer, but if you want a higher audio level, you can use a transistorized output stage to drive a loudspeaker. Several output stages are suggested within this book.

The electronic organ will play only one note at a time, but the number of notes is unlimited. Thus, different tone can be selected by touching the probe onto different keyboard contacts, so the circuit functions as a musical instrument that is played via the touch probes.

The tonal range is given by C1, which can be altered in a large range of values. Values between 0.022 and 1 µF can be used experimentally.

A schematic diagram of the Electronic Organ is given in Fig. 8.15.

Parts List: Electronic Organ

IC1	4093 CMOS integrated circuit
X1	Piezoelectric transducer, Radio Shack 273-073 or equivalent
P	Probe
S1	SPST toggle or slide switch
B1	6 or 9 V, four AA cells or battery
R1	10,000 Ω, 1/4 W, 5% resistor
R2 to Rn+1	100,000 to 1,000,000 Ω trimmer potentiometers
C1	0.022 µF ceramic or metal film capacitor
C2	100 µF, 12 WVDC electrolytic capacitor

The circuit can be powered from AA cells or a 9 V battery. Powerful output stages can be used to drive small loudspeakers, as suggested in other projects. Each note is adjusted by the corresponding trimmer potentiometer. Current drain is only few milliamperes, extending battery life.

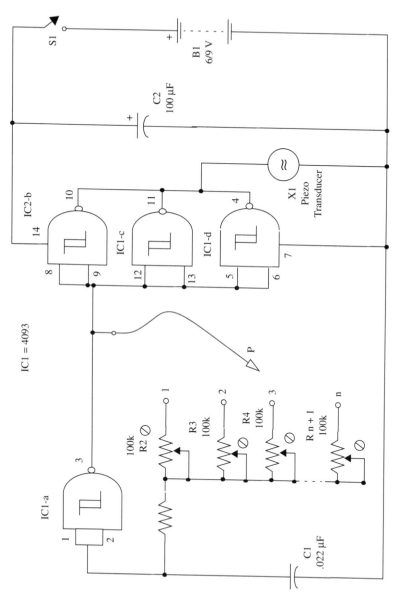

Figure 8.15   Electronic Organ.

Project 121: Electronic Organ with Vibrato (P) (E)

A low-frequency oscillator added to the Electronic Organ (Project 120) will produce a vibrato effect. The produced notes will be modulated in a frequency that is controlled by R1. Depth is controlled by R4.

As in Project 120, the notes are produced by touching the probe on the keyboard. The transducer is a piezoelectric unit, but if you want higher audio levels, you can use a transistorized output stage driving a loudspeaker. Other projects in this book show how to make the necessary modifications.

The electronic organ is powered from four AA cells or a battery, and current drain is very low, ranging from 0.5 mA when the tone is off to 5 mA when the tone is on.

A schematic diagram of the Electronic Organ with Vibrato is shown in Fig. 8.16.

Parts List: Electronic Organ with Vibrato

IC1	4093 CMOS integrated circuit
X1	Piezoelectric transducer, Radio Shack 273-073 or equivalent
S1	SPST toggle or slide switch
B1	6 or 9 V, four AA cells or battery
P	Probe (see text)
R1	2,200,000 Ω potentiometer or trimmer potentiometer
R2	100,000 Ω, 1/4 W, 5% resistor
R3	47,000 Ω, 1/4 W, 5% resistor
R4	100,000 Ω potentiometer or trimmer potentiometer
R5	47,000 Ω, 1/4 W, 5% resistor
R6	10,000 Ω, 1/4 W, 5% resistor
R7 to Rn+6	Trimmer potentiometers, 100,000 Ω
C1	0.22 μF or 0.47 μF ceramic or metal film capacitor
C2	0.1 to 0.47 μF ceramic or metal film capacitor
C3	0.022 μF ceramic or metal film capacitor
C4	100 μF, 16 WVDC electrolytic capacitor

The keyboard is made as described in Project 120. C1 and C2 can be altered to change the vibrato effect. If you want more than seven notes to be produced by the organ, use 470 kΩ or 1 MΩ units.

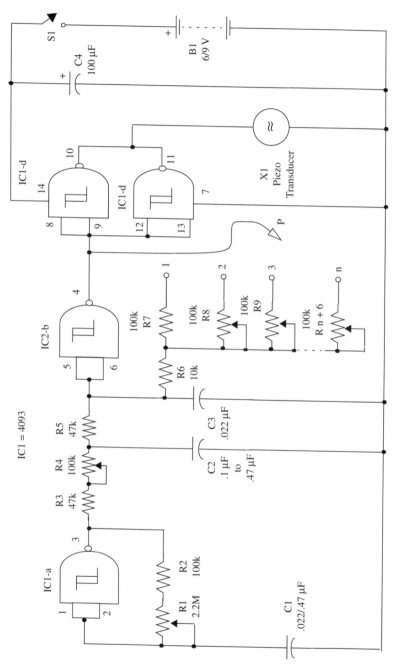

Figure 8.16 Electronic Organ with Vibrato.

Project 122: Sound-Activated LED (E)

Any sound received by the microphone will produce a flash in the output LED. This project can be altered to drive a relay and control external loads from sound sources.

Sensitivity is controlled by R1, and you can use small dynamic microphones with impedances ranging from 4 to 200 Ω. A small loudspeaker can be used as a sensitive microphone in this experimental project. The impedance of microphones can be increased by using a small transistorized output transformer. The circuit can also be used to detect vibrations or mechanical shocks. In this case, the transducer can be a loudspeaker.

The circuit is powered from AA cells or a battery, and current drain is very low when the LED is off.

IC1 is a JFET operational amplifier (op amp) that drives the 4093. The gates of the 4093 are used as inverters (IC1-a) and buffers (IC1-b, c, and d). To drive a relay, you should use a transistorized output stage as described in several other projects in this book.

A schematic diagram of the Sound-Activated LED is shown in Fig. 8.17.

Proper positioning of the polarized components (diode, LED, and electrolytic capacitor) must be observed.

Parts List: Sound-Activated LED

IC1	CA3140 JFET operational amplifier
IC2	4093 CMOS integrated circuit
LED1	Red common LED
D1	1N4148 general purpose silicon diode
MIC	4 to 200 Ω dynamic microphone or small loudspeaker (see text)
S1	SPST toggle or slide switch
B1	6 or 9 V, four AA cells or battery
R1	1,000,000 Ω potentiometer
R2	100,000 Ω, 1/4 W, 5% resistor
R3	1,000,000 Ω, 1/4 W, 5% resistor
R4	1,000 Ω, 1/4 W, 5% resistor
C1	0.01 µF ceramic or metal film capacitor
C2	100 µF, 12 WVDC electrolytic capacitor

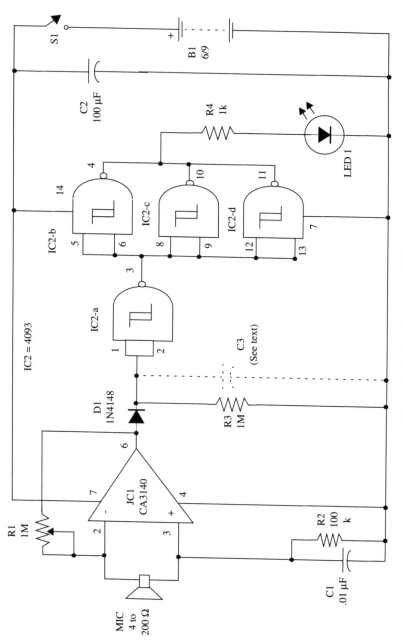

Figure 8.17   Sound-Activated LED.

Capacitor C3 adds inertia to the circuit. Using large values (between 1 and 10 µF), any short duration sound will turn on the LED for a time period ranging from some seconds to some minutes. A capacitor (C3) in this position is indicated if you want to modify the circuit to drive a relay.

C1 controls the frequency response. R1 adjusts gain and therefore the sensitivity of the circuit. The microphone can be installed at a distance from the circuit, and you can use common wires.

Project 123: Simple Bargraph (E)

This circuit drives four LEDs in response to an analog voltage and can be used to provide a visual indication of the instantaneous power being developed by an audio amplifier.

You can extend the project by using two chips, and you will be able to continuously monitor the audio output power of both sides of your stereo system. This will permit to you properly set the balance control for equalized outputs.

Transformer T1, in the input, isolates the circuit from the amplifier, providing the necessary safety from shorts and shocks.

The circuit can be used with amplifiers ranging from 0.1 to 100 W. Rx is chosen according the output power of the amplifier. Values are giving in Table 8.1.

Table 8.1   Amplifier Output vs. Rx

Amplifier output power (W)	Rx ($\Omega \times$ W)
0.1 to 1	–
1 to 5	10 $\Omega \times$ 0.5 W
5 to 20	22 $\Omega \times$ 1 W
20 to 50	47 $\Omega \times$ 1 W
50 to 100	100 $\Omega \times$ 1 W

Resistors R3, R4, R5, and R6 determine the dynamic range in which the circuit operates. Variations in these components can be made to match the action of the LEDs with logarithmic or other scales.

R1 adjusts sensitivity and, depending on the application, you can set the value to a fixed point using a trimmer potentiometer.

The circuit can be powered from power supplies ranging from 6 to 12 V. Resistors R7 to R10 determine the light level of the LEDs and can be altered according to the power supply voltage.

A schematic diagram of the Simple Bargraph is given in Fig. 8.18.

Audio signals come from the loudspeaker output in the amplifier. Any small transformer with a 117 Vac primary, a secondary rated from 5 to 12.6 V, and currents between 100 and 500 mA can be used in this project.

Proper positioning of the polarized components (LEDs, diode, and electrolytic capacitor) must be observed.

R1 adjusts sensitivity, and R6 adjusts the threshold of operation (i.e., when the first LED turns on).

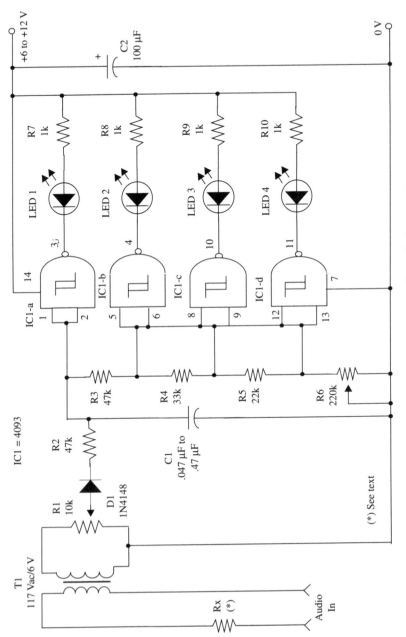

Figure 8.18   Simple Bargraph. Rx depends on the input audio power.

Parts List: Simple Bargraph

IC1	4093 CMOS integrated circuit
D1	1N4148 general purpose silicon diode
LED1 to LED4	Red common LEDs
T1	Transformer, 117 Vac to approx. 5 to 12 V, 150 mA to 500 mA (Radio Shack 273-1385, 12.6 V, 300 mA is suitable for this project—see text)
Rx	Value according amplifier output power (see Table 8.1)
R1	10,000 Ω potentiometer
R2	47,000 Ω, 1/4 W, 5% resistor
R3	47,000 Ω, 1/4 W, 5% resistor
R4	33,000 Ω, 1/4 W, 5% resistor
R5	22,000 Ω, 1/4 W, 5% resistor
R6	220,000 Ω potentiometer or trimmer potentiometer
R7 to R10	1,000 Ω, 1/4 W, 5% resistors
C1	0.047 to 0.47 µF ceramic or metal film capacitor (see text)
C2	100 µF, 16 WVDC electrolytic capacitor

Bass response is given by C1. You can vary this component in the range given in the schematic diagram to achieve the best performance.

Two units can be mounted to monitor a stereo system. Each unit is then wired to a channel of the stereo amplifier.

Project 124: General Purpose Automatic Switch (E) (P)

This circuit turns a load on and off automatically at a preset rate. You can use it to control warning lamps, in decorations, and to switch sirens, motors, home appliances, and so forth on and off.

On and off intervals can be adjusted from a few seconds to several minutes, according the intended application, by varying the value of C1. With values between 0.22 and 0.47 µF, we have time intervals that can be adjusted from a fraction of a second to a few seconds. Using a 100 µF capacitor, we get time intervals ranging from 1 or 2 minutes to 7 or 8 minutes.

Current requirements of the load are limited by the relay's contacts. You can both use a 1 A DPDT 12 V mini relay (Radio Shack 275-249) or a 10 A SPDT mini (Radio Shack 275-248). Relays with coils rated from 6 V can also be used in this project. This, of course, reduces the power supply voltage to the same value.

The circuit can be powered from battery or ac-to-dc converters. Current requirements are basically determined by the relay.

A schematic diagram of the General Purpose Automatic Switch is given in Fig. 8.19.

Parts List: General Purpose Automatic Switch

IC1	4093 CMOS integrated circuit
Q1	2N2222 NPN General purpose silicon transistor
D1	1N4148 general purpose silicon diode
K1	6 or 12 V relay (see text)
R1	2,200,000 Ω potentiometer
R2	100,000 Ω, 1/4 W, 5% resistor
R3	2,200 Ω, 1/4 W, 5% resistor
C1	0.22 µF to 100 µF ceramic, metal film or electrolytic capacitor (see text)
C2	100 µF, 16 WVDC electrolytic capacitor

This layout uses a DPDT relay, but you can change it to accommodate a different relay.

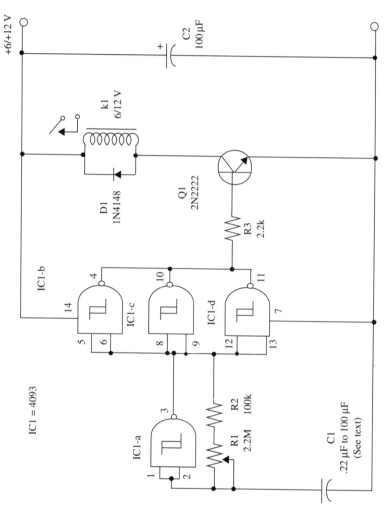

Figure 8.19 General Purpose Automatic Switch.

Project 125: Dark-Activated Flasher (P)

You can use this circuit to turn on a warning lamp automatically at dusk and off at dawn, as described in the LDR version (see Project 101).

This circuit uses a photo transistor as the sensor, and operation is the same as described in Project 101 except for the low-frequency oscillator and the transducer.

R1 adjusts the turn-on light level, and R2 determines the frequency of the flashes.

The lamp (or many lamps) can be controlled by the relay, but you can also use the circuit to control other loads such as appliances, motors, solenoids, and so forth. The frequency can be altered by changing C1. See Project 124 for relevant information.

A schematic diagram of the Dark-Activated Flasher is shown in Fig. 8.20.

Parts List: Dark-Activated Flasher

IC1	4093 CMOS integrated circuit
Q1	TIL81 or equivalent photo transistor
D1	1N4148 general purpose silicon diode
Q2	2N2907 PNP general purpose silicon transistor
K1	6 or 12 V relay (see text)
R1, R2	2,200,000 Ω potentiometers
R3	100,000 Ω, 1/4 W, 5% resistor
R4	2,200 Ω, 1/4 W, 5% resistor
C1	0.22 µF to 100 µF ceramic, metal film, or electrolytic capacitor (see text)
C2	100 µF, 16 WVDC electrolytic capacitor

The layout can be altered to suit the relay used. Any phototransistor can be used in this circuit. Even a common power transistor such as the 2N3055, without its cover, can act as a sensitive phototransistor.

C1 determines the turn-on and turn-off rate as required by the intended application. See Project 124 for more information about this component.

The relay depends on the controlled load. A mini DPDT 1 A relay (Radio Shack 275-249) can be used to control small appliances and lamps up to 100 W.

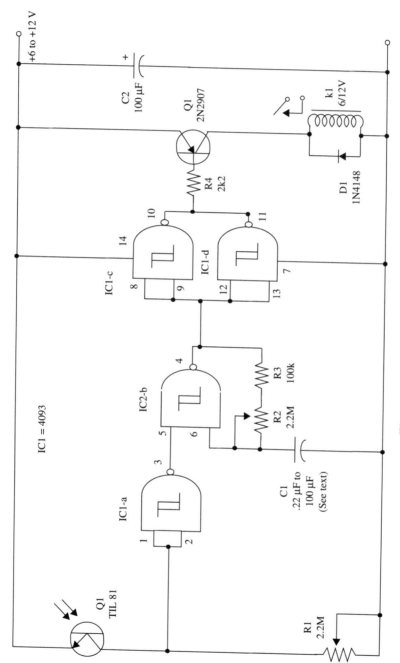

Figure 8.20 Dark-Activated Flasher

Project 126: Touch Switch I (E)

Small appliances, lamps, and motors can be controlled by a simple touch with this experimental circuit. Loads are on only while the sensor is touched. The sensor is a small plate that should not be placed far from the circuit.

The circuit can be powered from 6 or 12 V supplies, depending on the relay. Do not use transformerless power supplies in this project, as they may lead to shock or dangerous situations. Sensitivity can be increased by connecting point X to a good earth.

Cx is used if the circuit tends toward erratic operation due to signals picked up by the sensor or its wire.

A schematic diagram of the Touch Switch is shown in Fig. 8.21.

Parts List: Touch Switch I

IC1	4093 CMOS integrated circuit
X1	sensor plates (see text)
Q1	2N2222 NPN general purpose silicon transistor
D1	1N4148 general purpose silicon transistor
K1	6 or 12 V relay (see text)
R1	100,000 Ω, 1/4 W, 5% resistor
R2	10,000,000 Ω, 1/4 W, 5% resistor
R3	4,700 Ω, 1/4 W, 5% resistor
Cx	See text
C1	100 μF, 16 WVDC electrolytic capacitor

The touch sensor is a 2 × 2 inch metal plates and is connected to the circuit by a 4 to 10 inch common wire. Sensitivity can be changed by altering R2. Values between 2.3 and 10 MΩ can be used experimentally.

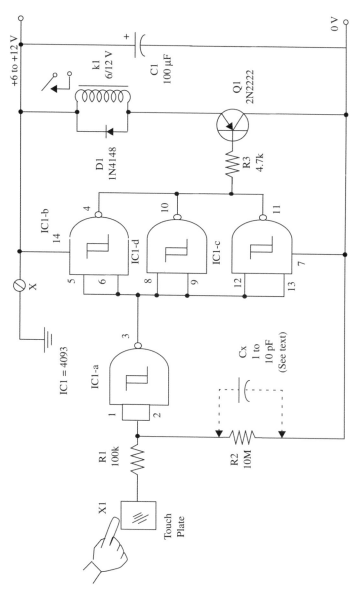

Figure 8.21   Touch Switch I

Project 127: Touch Switch II (E)

This circuit allows you to control loads up to 3 A (9 to 12 V) by touching a metal plate. The circuit can be used to control small dc motors, solenoids, lamps, and other small dc appliances.

The circuit is a modification of the Project 126, where the relay is replaced by an SCR. We can power the circuit from 9 to 12 V supplies. As the SCR is subjected to a 2 V voltage when in operation (on), the load receives the power supply voltage less 2 V.

D1 is used only with inductive loads such as motors and solenoids. R1 determines sensitivity and can be varied in a range between 2.2 and 10 MΩ. You can replace this resistor by a 10 MΩ potentiometer for sensitivity control.

A schematic diagram of the Touch Switch II is shown in Fig. 8.22.

Parts List: Touch Switch II

IC1	4093 CMOS integrated circuit
SCR	TIC106, C106, MCR106, or any low-power silicon controlled rectifier
D1	1N4148 general purpose silicon diode
X1	Touch sensor (see text)
R1	10,000,000 Ω, 1/4 W, 5% resistor
R2, R3	4,700 Ω, 1/4 W, 5% resistor
S1	SPST momentary switch
C1	100 µF, 16 WVDC electrolytic capacitor

The touch switch uses two metal plates that are touched simultaneously with your fingers. Sensitivity is determined by R1, which can be altered in the range between 2.2 and 10 MΩ according the touch switch wires. Long wires need low values to reduce the effect of picked-up signals that can cause erratic operation of the circuit.

The SCR must be mounted on a heatsink. Remember that this is a self-latching circuit. A single touch in the sensor turns the load on until S1 is pressed to turn the circuit off.

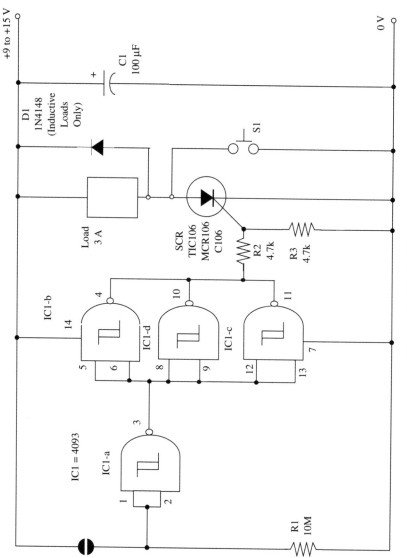

Figure 8.22 Touch Switch II.

Project 128: Touch Switch III (E)

This circuit can drive a load up to 2 A without the use of a relay. You can use it to control small dc motors, lamps, solenoids, and heaters, as described in Project 127.

The circuit replaces the relay with a Darlington power transistor, and it can be powered from 6 to 12 V supplies. Current drain depends on the controlled load.

A schematic diagram of the Touch Switch III is shown in Fig. 8.23.

Parts List: Touch Switch III

IC1	4093 CMOS integrated circuit
R1	100,000 Ω, 1/4 W, 5%
R2	4,700,000 Ω, 1/4 W, 5%
R3	4,700 Ω, 1/4 W, 5%
Q1	TIP115 NPN Darlington power transistor
D1	1N4148 general purpose silicon diode
X1	Touch sensor (see text)
R1	4,700,000 Ω to 10,000,000 Ω, 1/4 W, 5% resistor (see text)
R2	4,700 Ω, 1/4 W, 5% resistor
C1	100 µF, 16 WVDC electrolytic capacitor

The transistor must be mounted on a heatsink. Note that this is a non-latching circuit. The loads are on only during the time you are touching the sensor plates.

You can also replace the bipolar transistor with a power FET to increase power capabilities of the circuit. Proper positioning of the load must be observed if it is a motor or other polarized device.

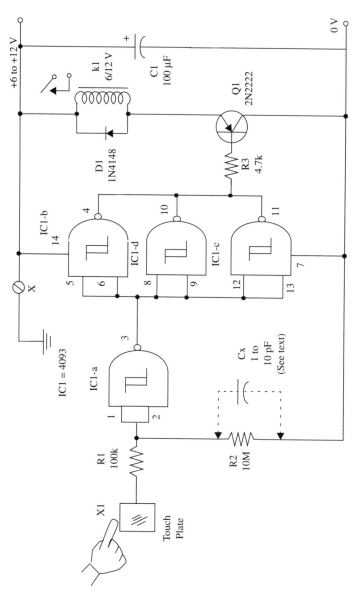

Figure 8.23 Touch Switch III.

Project 129: Opto-Coupler Interface (P) (E)

Using this circuit, you can control home appliances, motors, lamps, heaters, and several other ac or dc loads from a computer or other digital equipment. The opto-coupler isolates the load from the control circuit, providing a high level of security to the user.

The circuit can control loads up to 1 A with a mini relay (Radio Shack 275-249), or heavy-duty loads with a 10 A mini relay (Radio Shack 275-2488).

The circuit turns on the load with a high logic level at the input, and Rx is used to limit the current flow through the LED, according the digital output signal. The table within the schematic diagram shows Rx values as function of output voltage (V_{cc}). The circuit can be powered from 6 to 12 V supplies, depending on the relay.

A schematic diagram of the Opto-Coupler Interface is shown in Fig. 8.24.

Parts List: Opto-Coupler Interface

IC1	4093 CMOS integrated circuit
IC2	4N25 or equivalent opto-coupler
D1	1N4148 general purpose silicon diode
Q1	2N2907 PNP general purpose silicon transistor
K1	6 or 12 V relay (see text)
Rx	See text
R1	100,000 Ω, 1/4 W, 5% resistor
R2	4,700 Ω, 1/4 W, 5% resistor
C1	100 µF, 16 WVDC electrolytic capacitor

The circuit can be mounted on solderless boards or printed circuit boards, and layout depends on the relay employed and the size and shape of other components. Proper polarity of the input wires must be observed.

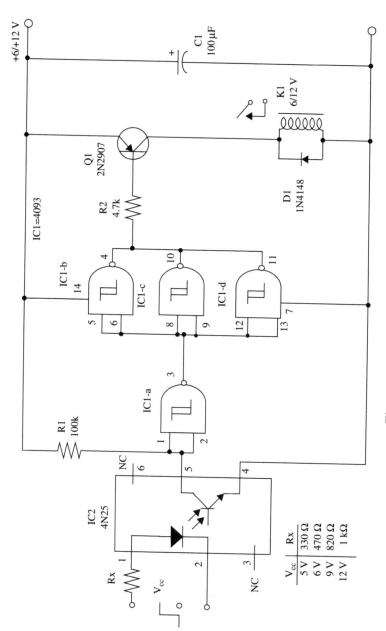

Figure 8.24 Opto-Coupler Interface.

Project 130: Twin-Lamp AC Flasher (P)

This circuit employs two lamps that will flash at a frequency rate determined by R1. The circuit operation is such that lamp 1 (L1) is off when the lamp 2 (L2) is on, and vice versa.

Note that this is a half-wave circuit, so the lamps will flash with half the rated power. This factor should be considered in determining appropriate circuit applications, which include warning systems, decoration, and so on.

The frequency range can be altered by changing C2 values. Values can range from 0.22 to 10 µF. The circuit can easily drive two 200 W lamps. The SCRs must be mounted on heatsinks.

A schematic diagram of the Twin-Lamp AC Flasher is shown in Fig. 8.25.

Parts List: Twin-Lamp AC Flasher

IC1	4093 CMOS integrated circuit
SCR1, SCR2	TIC106, MCR106, or C106 SCR, 200 V peak inverse voltage (PIV)
D1, D2	1N4002 or equivalent (50 V, 1 A) silicon diodes
T1	6.3 V, 300 mA CT transformer, primary 117 Vac (see text)
F1	1 to 5 A fuse, depending on the lamp
S1	SPST toggle or slide switch
L1, L2	5 to 200 W incandescent lamp, 117 V
R1	2,200,000 Ω potentiometer
R2	100,000 Ω, 1/4 W, 5% resistor
R3, R5	4,700 Ω, 1/4 W, 5% resistors
R4, R6	2,200 Ω, 1/4 W, 5% resistors
C1	1,000 µF, 25 WVDC electrolytic capacitor
C2	0.22 µF to 0.47 µF ceramic or metal film capacitor

Special care should be taken with the high-current and high-voltage lines. The circuit can be housed in a plastic box with all the high voltage lines completely isolated from external touch.

Proper positioning of the polarized components as the diodes, electrolytic capacitors and the SCRs must be observed. SCRs rated to 200 peak inverse voltage (PIV) or more can be used in this project.

The fuse is chosen according the lamps used. An 1 A fuse is employed with 5 to 40 W lamps, a 2 A with 60 to 100 W lamps, and a 5 A fuse with lamps rated from 120 to 200 W.

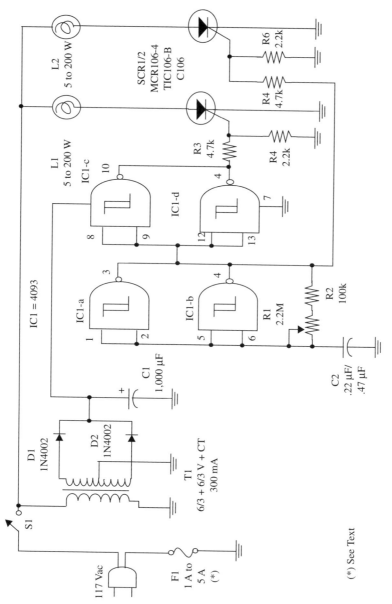

Figure 8.25 Twin-Lamp AC Flasher.

Project 131: Flickering-Flame Effect (P) (E)

This circuit simulates a campfire or candle by varying the brightness of a lamp. You can use it in fireplaces or inside antique oil lamps to obtain a realistic simulation of their normal operation.

The circuit can drive 5 to 200 W incandescent lamps, and the effect is adjusted by R1.

A schematic diagram of the Flickering-Flame Effect is shown in Fig. 8.26.

Parts List: Flickering-Flame Effect

IC1	4093 CMOS integrated circuit
SCR	TIC106, MCR106, or C106 (200 PIV) silicon controlled rectifier
D1, D2	1N4002 silicon rectifier
T1	6.3 V, CT, 300 mA transformer, primary 117 Vac
F1	5 A fuse with holder
S1	SPST toggle or slide switch
R1	1,000,000 Ω potentiometer
R2	330,000 Ω, 1/4 W, 5% resistor
R3	4,700 Ω, 1/4 W, 5% resistor
R4	1,000 Ω, 1/4 W, 5% resistor
C1	1,000 µF, 25 WVDC electrolytic capacitor
C2	0.15 µF ceramic or metal film capacitor
L1	5 to 200 W incandescent lamp (117 Vac)

The SCR must be mounted on a heatsink. Proper positioning of the polarized components must be observed.

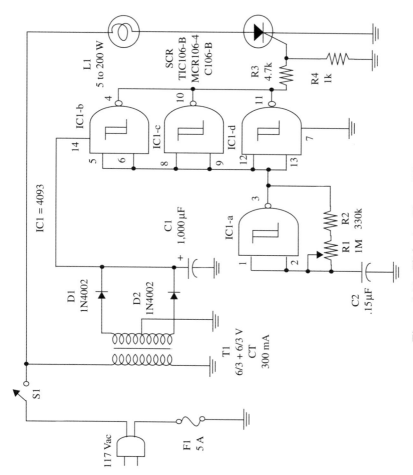

Figure 8.26   Flickering-Flame Effect.

Project 132: Dark-Activated Lamp (P)

You can use this circuit to activate an incandescent lamp when it gets dark and switch it off again when it gets light.

The circuit replaces a relay with an SCR and can drive lamps up to 200 W. Note that the SCR is a half-wave device, so the incandescent lamp glows with half of its normal brightness.

Power comes from the ac power line, and sensitivity is adjusted by R1. The sensor is an LDR that should be placed as far as possible from the controlled lamp to avoid feedback. The LDR should receive only ambient or natural light.

Any small transformer rated from 100 to 500 mA can be used in this project. An LED to indicate power on is optional.

A schematic diagram of the Dark-Activated Lamp is shown in Fig. 8.27.

Parts List: Dark-Activated Lamp

IC1	4093 CMOS integrated circuit
SCR	TIC106, MCR106, or C106 (200 PIV) silicon controlled rectifier
D1, D2	1N4002 or equivalent silicon rectifiers
LDR	CdS photocell, Radio Shack 276-1657
S1	SPST toggle or slide switch
F1	5 A fuse and holder
T1	6.3 V, 300 mA, CT transformer
L1	5 to 200 W, 117 V incandescent lamp
R1	1,000,000 Ω potentiometer
R2	10,000 Ω, 1/4 W, 5% resistor
R3	4,700 Ω, 1/4 W, 5% resistor
R4	1,000 Ω, 1/4 W, 5% resistor
C1	1,000 μF, 25 WVDC electrolytic capacitor
C2	0.1 μF ceramic or metal film capacitor

A 200 peak inverse voltage (PIV) SCR is used in this circuit. This SCR must be mounted on a heatsink. Take care with the high voltage lines, using appropriate wires and avoiding any contact with external parts of the device.

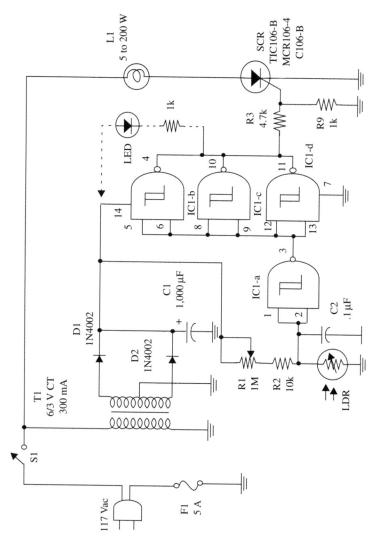

Figure 8.27 Dark-Activated Lamp.

The circuit can be housed into a plastic box, and the LDR is placed at a distance from the lamp to receive only the ambient light. Adjustment of the sensitivity level is made by R1 to turn on the lamp at dusk.

C1 values can be varied to obtain greater immunity to short flashes of light that occur during electric storms and can cause unstable operation of the circuit.

Project 133: Power Bistable with Magnetic Switches (P)

You can turn heavy-duty loads on and off from the ac power line using the low-current magnetic switches in this circuit. The project can used as part of alarms, automatic mechanisms, or as a part of more complex configurations.

When the sensor K1 is activated, the bistable formed by IC1-a and b acts, and the SCR is turned on. Then, the lamp remains on until the bistable changes its state again by the action of X2.

The circuit is a half-wave control, because we are using an SCR, but you can easily alter it to operate as a full-wave control by putting a full-wave bridge on the ac power line input.

The SCR is a 200 peak inverse voltage (PIV) device that must be mounted on a heatsink. Lamps or heaters (resistive loads) can be controlled in the range between 5 and 200 W.

A schematic diagram of the Power Bistable with Magnetic Switches is shown in Fig. 8.28.

Parts List: Power Bistable with Magnetic Switches

IC1	4093 CMOS integrated circuit
SCR	TIC106, C106 or MCR106 200 PIV silicon controlled rectifier
D1, D2	1N4002 or equivalent silicon rectifiers
X1, X2	Reed switches
L1	5 to 200 W 117 V incandescent lamp
F1	5 A fuse and holder
S1	SPST toggle or slide switch
T1	6.3 V, CT, 300 to 500 mA transformer
R1, R2	100,000 Ω, 1/4 W, 5% resistors
R3	4,700 Ω, 1/4 W, 5% resistor
R4	1,000 Ω, 1/4 W, 5% resistor
C1	1,000 μF, 25 WVDC electrolytic capacitor

Be careful with the wires to the sensor. These wires are not isolated from the ac power line.

Positions of the polarized components (diodes, electrolytic capacitors, SCRs) must be observed.

Erratic operation of the circuit, in case of long wires to the sensors, can be reduced by reducing the values of R1 and R2. You can use resistor values as low as 10 kΩ in this function.

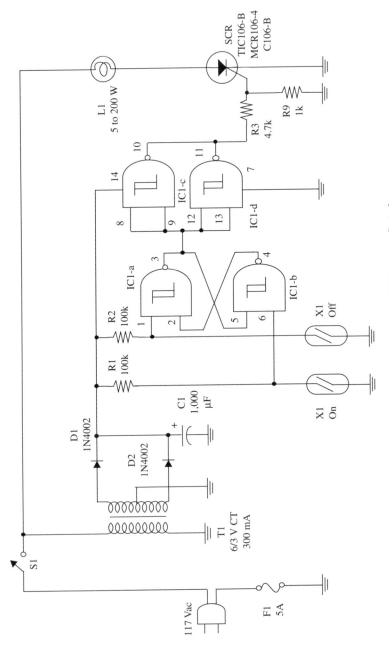

Figure 8.28 Power Bistable with Magnetic Switch.es

Project 134: AC Lamp Flasher (P)

This project can be used as visual alarm or in commercial advertising. The circuit produces short flashes using an ac incandescent lamp.

The duration of the flashes is given by C2 and R3, and the frequency or repetition rate is adjusted by R1. C1 can be varied to change the range of the flash rate. Values between 100 and 1,000 µF can be used experimentally according the intended application.

The circuit consists of a variable duty cycle oscillator that drives an SCR. As its load, the SCR has a common incandescent lamp of from 5 to 200 W. Remember that SCRs are half-wave devices that drive the loads with half of the total power.

A schematic diagram of the AC Lamp Flasher is given in Fig. 8.29.

Parts List: AC Lamp Flasher

IC1	4093 CMOS integrated circuit
SCR	TIC106, MCR106, or C106 200 V PIV silicon controlled rectifier
D1, D2	1N4002 or equivalent silicon rectifiers
T1	6.3 V, CT, 300 to 500 mA transformer
F1	5 A fuse and holder
S1	SPST toggle or slide switch
R1	1,000,000 Ω potentiometer
R2	10,000 Ω, 1/4 W, 5% resistor
R3	47,000 Ω, 1/4 W, 5% resistor
R4	4,700 Ω, 1/4 W, 5% resistor
R5	1,000 Ω, 1/4 W, 5% resistor
C1	100 µF, 16 WVDC electrolytic capacitor
C2	10 µF, 16 WVDC electrolytic capacitor
C3	1,000 µF, 15 WVDC electrolytic capacitor
L1	5 to 200 W incandescent lamp, 117 V

The SCR must be mounted on a heatsink. Any small transformer with its secondary rated from 300 to 500 mA can be used.

You can also alter this circuit for a full-wave control by inserting a full-wave bridge (200 V, 4 A diodes) in the load line. The flashes are adjusted by R1.

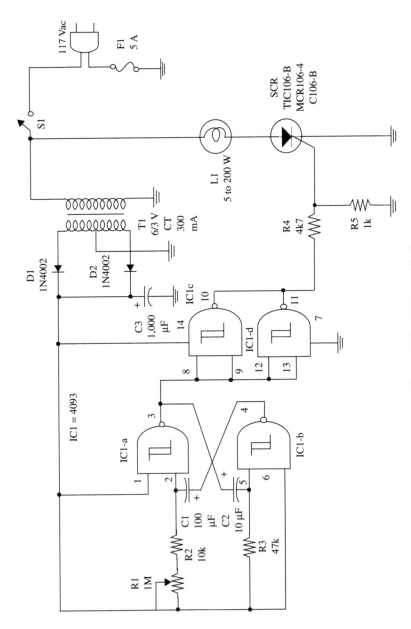

Figure 8.29   AC Lamp Flasher

Project 135: Full-Wave Touch Switch for AC Loads (P)

AC loads up to 800 W can be controlled by touch with this circuit. The project is completely isolated from the ac power line, avoiding dangerous shocks.

A microampere dc current through your fingers can control heavy-duty loads such as home appliances, motors, and heaters up to 8,800 W.

The circuit consists of a bistable touch-controlled multivibrator that drives a triac through a unijunction transistor (UJT) and an isolating transformer. The triac, a TIC226-B, is rated for loads up to 800 W (8 A, 117 V) and must be mounted on a heatsink.

Wires to the touch plates can be as long as you want, and you can place it some distance from the circuit to be controlled. As the control line operates with a low voltage and low current, no special care needs to be taken with the connection.

A schematic diagram of the Full-Wave Touch Switch for AC Loads is given in Fig. 8.30.

Parts List: Full-Wave Touch Switch for AC Loads

IC1	4093 CMOS integrated circuit
IC2	7812 voltage regulator IC
Triac	TIC226-B, Texas Instruments
Q1	2N2646 unijunction transistor
D1, D2	1N4002 or equivalent silicon rectifiers
S1	SPST toggle or slide switch
X1, X2	Touch sensor (see text)
F1	10 A fuse and holder
T1	12.6 V, CT, 300 to 500 mA transformer
T2	1:1 pulse transformer
R1, R2	10,000,000 Ω, 1/4 W, 5% resistors
R3	100,000 Ω trimmer potentiometer
R4	10,000 Ω, 1/4 W, 5% resistor
C1	1,000 μF, 25 WVDC electrolytic capacitor
C2	0.047 μF ceramic or metal film capacitor
C3	100 μF, 16 WVDC electrolytic capacitor

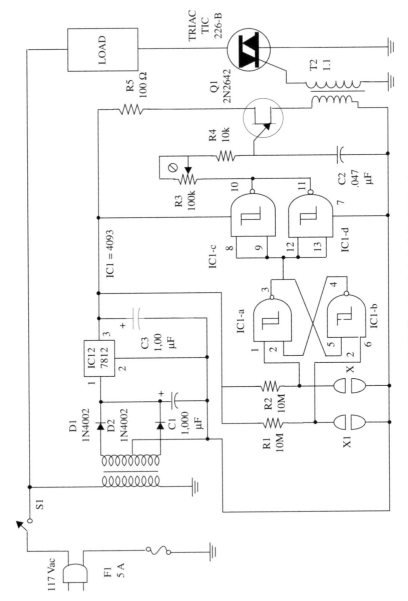

Figure 8.30 Full-Wave Touch Switch for AC Loads

Make sure that the ac high current line isn't connected to the board and that the wire used is appropriate for this task. The triac must be mounted on a large heatsink.

T2 is a pulse transformer (1:1) as used in many circuits to trigger SCRs and triacs. X1 and X2 are touch switches made with two small metal plates that are simultaneously touched by the fingers. R3 is adjusted to the maximum power of the load when the circuit is on.